Water Governance, Policy and Knowledge Transfer

In an increasingly global community of researchers and practitioners, new technologies and communication means have made the transfer of policies from one country or region to another progressively more prevalent. There has been a lot of attention in the field of public administration paid to policy transfer and institutional transplantation. This book aims to create a better understanding of such transfers in the water management sector. These include the adoption of modern water management concepts, such as integrated water resources management and forms of water governance, which are strongly promoted and sometimes also imposed by various international organizations. Transfers also occur within the scope of development aid or for the purpose of creating business opportunities. In addition, many research organizations, consultancies and governmental agencies are involved in cross-border work.

The purpose of this book is therefore to present practical examples of the transfer of modern water management from one locality to another and to critically discuss the transferability of policy and governance concepts by analysing the contextual needs and factors. Case studies are included from North America, Europe, the Middle East and Asia. It is argued that in many cases context matters in water management and that there is no panacea or universal concept that can be applied to all countries or regions with different political, economic, cultural and technological contexts. Yet it is also shown that some countries are facing pressing and similar water management issues that cut across national borders, and hence the transfer of knowledge may be beneficial.

Cheryl de Boer is a Post-Doctoral researcher at the Centre for Studies in Technology and Sustainable Development (CSTM), University of Twente, the Netherlands, as well as Coordinator of the Twente Water Centre.

Joanne Vinke-de Kruijf just finalized her PhD at CSTM and at the Water Engineering and Management Department of the University of Twente, the Netherlands and is currently international project manager at regional water authority Regge en Dinkel.

Gül Özerol is a researcher and PhD candidate at CSTM, University of Twente, the Netherlands.

Hans Bressers is a Professor of Policy Studies and Environmental Policy and founder of CSTM, University of Twente, the Netherlands, as well as Chair of the Twente Water Centre.

Earthscan Studies in Water Resource Management

Water Management, Food Security and Sustainable Agriculture in Developing Economies
Edited by M. Dinesh Kumar, M.V.K. Sivamohan and Nitin Bassi

Governing International Watercourses
River Basin Organizations and the Sustainable Governance of Internationally Shared Rivers and Lakes
By Susanne Schmeier

Transferable Groundwater Rights
Integrating Hydrogeology, Law and Economics
By Andreas N. Charalambous

Contemporary Water Governance in the Global South
Scarcity, Marketization and Participation
Edited by Leila Harris, Jacqueline Goldin and Christopher Sneddon

For more information and to view forthcoming titles in this series, please visit the Routledge website: http://www.routledge.com/books/series/ECWRM/

Water Governance, Policy and Knowledge Transfer

International studies on contextual water management

Edited by
Cheryl de Boer, Joanne Vinke-de Kruijf, Gül Özerol and Hans Bressers

Routledge
Taylor & Francis Group
LONDON AND NEW YORK

earthscan
from Routledge

First published 2013 by Routledge

Published 2019 by Routledge
2 Park Square, Milton Park, Abingdon, Oxon OX14 4RN
52 Vanderbilt Avenue, New York, NY 10017

Routledge is an imprint of the Taylor & Francis Group, an informa business

British Library Cataloguing in Publication Data
A catalogue record for this book is available from the British Library

Library of Congress Cataloging-in-Publication Data
Water governance, policy and knowledge transfer : international studies in contextual water management / edited by Cheryl de Boer, Joanne Vinke-de Kruijf, Gül Özerol and Hans Bressers.
pages cm. – (Earthscan studies in water resource management)
Includes bibliographical references and index.
1. Water resources development. 2. Water resources development–Government policy. 3. Technology transfer. I. Boer, Cheryl de.
HD1691.W32325 2013
333.91–dc23
2012049457

ISBN13: 978-0-415-62597-5 (hbk)
ISBN13: 978-1-138-57389-5 (pbk)

Typeset in Baskerville
by Saxon Graphics Ltd, Derby

Contents

List of illustrations

Figures

Tables

Boxes

Contributors

Denie C.M. Augustijn is associate professor at the department of Water Engineering and Management at the University of Twente. He has interest in a broad range of topics related to water management varying from a better understanding of physical, chemical and ecological processes in water systems to application of this knowledge in policy processes.

Maya van den Berg holds a Master of Arts from the University of Groningen. She is a junior researcher at CSTM at the University of Twente, where she has been involved in several externally funded research projects on climate change and sustainability development. She pursues her research interests in a PhD on adaptation policies in local governments.

Cheryl de Boer is a Post-Doctoral researcher at CSTM at the University of Twente. She is the Coordinator of the Twente Water Centre and has research interests in water governance, resilience and multi-functional land use as a means for sustainable resource management, climate mitigation and adaptation at the local and global scale. Most recently, she received her PhD (Cum Laude) in governance and sustainability at the University of Twente.

Hans Bressers is Professor of Policy Studies and Environmental Policy at CSTM at the University of Twente. He is a member of the Committee on Sustainable Development of the Dutch Social Economic Council and of the national Advisory Committee on Water. At the University of Twente he is the Chair of the Twente Water Centre, that joins all water related research groups.

Cesar Casiano is General Director in NACE Sustainable Alternatives and Lecturer at Tecnológico de Monterrey and Universidad Iberoamericana. Mr Casiano obtained a MSc degree in Sustainable Development Sciences from Tecnológico de Monterrey and studied his bachelor's programme of international relations at the Universidad de las Américas Puebla, where he was awarded as the best student of his generation. He has worked in a range of environmental research at the international level including at the Swiss Federal Institute of Technology.

Jaap Evers is lecturer river basin governance at UNESCO-IHE and works as a consultant for MetaMeta. His research interests are on interactive implementation processes and integrated governance of river basins. Within these issues he has a special interest in rural–urban interactions and synergies for the development of sustainable and productive socio-ecological systems.

María Laura Franco-García is senior researcher in private and public partnerships towards sustainability at CSTM at the University of Twente. She is the former director of the academic department and master's programme of sustainable development sciences in Tecnológico de Monterrey, Mexico. She obtained her PhD in Lyon, France on environmental sciences and environmental engineering by training in Mexico City.

Elena Carina Gutiérrez-Díaz is an associate founder of the Comisión de Cuenca Presa Guadalupe, A.C. and an environmental consultant. She is a biologist by training from the Universidad Nacional Autónoma de México and has a Master's degree in sustainable development sciences from Tecnológico de Monterrey, Mexico.

Lina Hendrawati Tan is food technologist at FeyeCon Development and Implementation B.V., the Netherlands. She obtained an MSc in environmental and energy management and an MSc in chemical engineering with specialization in water technology, both from the University of Twente, the Netherlands.

Marcia Hughes (PhD), Visiting Professor in sociology and interim director at the Center for Social Research at the University of Hartford, Connecticut has been providing programme development, research and evaluation services to nonprofit organizations, and government and community-based organizations for twenty years. Her work contributes to interdisciplinary, multi-level collaboratives for creating positive social change at the local, national and international levels. The focus of her research and evaluation is on how to translate knowledge of effective practices to real world contexts, and to inform policy and organizational change necessary for successful implementation and impact.

Mine Islar is a PhD candidate at LUCSUS, the University of Lund, Sweden. She has professional and academic backgrounds in water management, integrated assessment and knowledge formation through the strategic interactions of state, private companies, multilateral institutions and non-governmental organizations. Her research consists of topics such as securitization, hydro-politics and water governance.

Stefan Kuks is Professor of Innovation and Water Policy Implementation at CSTM at the University of Twente. He has been involved in various national and international water policy studies (e.g. www.euwareness.nl).

Besides his part time professorship, Stefan is full-time chairman (dyke reeve) of the regional public water authority Regge en Dinkel in the east of the Netherlands. He is also an elected member in the executive committee of the Dutch Association of Water Management Authorities.

Rob de Loë holds the University Research Chair in Water Policy and Governance at the University of Waterloo, and is Director of the multi-university Water Policy and Group (www.wpgg.ca). Previously he held the Canada Research Chair in Water Management at the University of Guelph. During the past two decades, Rob has written extensively about water security and related concerns such as source water protection, water allocation and climate change adaptation. His advice on water policy and governance concerns is widely sought by government and non-government organizations. He is a frequent speaker about water to audiences that include school children, the general public, and leaders in business and government.

Kris Lulofs is senior research associate at CSTM at the University of Twente. He specializes in policy strategies, instrumentation and governance, especially in the fields of corporate social responsibility, climate change and water management. He has (co-)edited several books and published a large number of articles, book chapters and reports. Currently, he leads the governance track in the Ecoshape-Building with Nature programme and is booster team member of the Water Governance Centre. He is a senior member of the Netherlands Institute of Government.

Irna van der Molen is Coordinator of the Centre for Risk management, Safety and Security; programme leader for the WOTRO Integrated Programme 'Conflict and Environment in North Lebanon' (both University of Twente); and Assistant Professor Water Management, Security and Governance at the Maastricht School of Management. Her research interests evolve around water and environmental management in conflict settings, or in contexts that are characterized by political instability and fragility.

Farhad Mukhtarov (PhD) is a researcher at the Biotechnology and Society section, Biotechnology Department, Delft University of Technology. He specializes in public policy and environmental governance. Farhad's research interests include theories of policy change, integrated resource management, policy translation and learning and discourse theory. He is currently working on biofuel policies in Brazil and previously worked on biodiversity conservation in Vietnam, and water governance in England, Turkey, Azerbaijan and Kazakhstan.

Dan Murray is a Post-Doctoral Research Associate at the Water Policy and Governance Group, located within the Department of Environmental and Resource Studies, University of Waterloo. His research focuses on water governance and water policy, with a particular emphasis on collaborative approaches to governance.

Katharine Owens received her PhD in sustainability and governance from the University of Twente and is currently assistant professor in the Department of Politics and Government at the University of Hartford, Connecticut. Her research interests include natural resource decision-making, climate change, wetlands and campus sustainability.

Aysun Özen Tacer is leading the sustainability practice in Deloitte Consulting Turkey. She has a civil engineering and business management background. She has conducted research on water management with special emphasis on time and risk perception dimensions of water management decisions at CSTM at the University of Twente.

Gül Özerol is a researcher and PhD candidate at CSTM at the University of Twente. She has professional and academic experience in industrial engineering, integrated assessment, decision science and organizational management. Her main research interests are natural resource governance, agricultural water management, transboundary water management, public participation and multi-criteria methods.

Ellen Skoczenski graduated with a combined BA in environmental studies and sociology from the University of Hartford, Connecticut. As an undergraduate student Ellen worked under the supervision of Dr Katharine Owens and Dr Marcia Hughes to study the effectiveness of international development efforts in rural villages in India and Kenya, utilizing the Contextual Interaction Theory. Ellen worked for a regional food bank in Philadelphia before landing her current position with the Coastal Mountains Land Trust in Camden, Maine.

Nora Stel is a research fellow at Maastricht School of Management and works on her PhD research at Utrecht University's Centre for Conflict Studies. Nora's PhD research concerns local governance interactions between Lebanese and Palestinian authorities in South Lebanon. Her current research interests and projects evolve around (post-)conflict politico-institutional development; non-state governance in hybrid political orders; state-building discourses and practices; and the theorization of political authority and legitimacy – with a focus on the Middle East.

Carmen Teodosiu is full professor in the Department of Environmental Engineering and Management, Faculty of Chemical Engineering and Environmental Protection of the Gheorghe Asachi Technical University of Iasi, Romania (TUIasi). She obtained her chemical engineering degree at TUIasi (1981), Post-graduate Diploma with distinction (1995) and Master of Science with distinction in environmental science and technology (1996), at the International Institute of Hydraulics and Environmental Engineering, UNESCO-IHE Delft, The Netherlands, and her PhD in Chemistry at TUIasi (1998). Prof. Teodosiu's research and teaching experience is in the fields of Water and Wastewater Treatment Processes, Integrated Water Resources Management, Environmental Management/Sustainable Consumption.

Vera Vikolainen is a Post-Doctoral researcher at CSTM at the University of Twente. She holds a Bachelor's degree in Business administration from St.-Petersburg State Polytechnic University and a Master's degree in European Studies from the University of Twente. She joined CSTM in 2007 where she worked on several projects in the field of policy implementation and water management. Her PhD research focuses on the feasibility of Building with Nature projects in the context of EU Natura 2000 implementation. Her research interests include EU environmental policy, water management and policy implementation.

Joanne Vinke-de Kruijf just finalized her PhD at CSTM and at the Water Engineering and Management Department of the University of Twente and is currently international project manager at regional water authority Regge en Dinkel. In her PhD research, she examines the effectiveness of projects that are implemented with Dutch financial support and expertise. In her research, she focuses on the human dimension of (international) water management. Her research interests include water governance, multi-actor processes, evaluation and international knowledge transfer.

Foreword

Doing better and delivering worse: pathology of water experts

Helen Ingram

Decades ago, Aaron Wildavsky (1977) called it political pathology to continue or increase support for policy solutions that turn out mediocre or poor performances. Just such an ailment currently afflicts water policy worldwide. Most studies, including those in this volume, conclude that a bundle of widely recommended and adopted policy transfers swept together under the rubric of Integrated Water Resources Management (IWRM) deliver results that fall far short of promise. In the last decade since the emergence of IWRM, hundreds of conferences promoting its embodiment in model statutes, participatory planning processes, and best practices and the transfer of professional knowledge have taken place. Yet, as the first chapter of this book notes, water problems continue unabated and even worsen. In response, water experts prescribe doing more of the same things and tacking on a growing list of additional but similar recommendations.

In contrast, this volume argues that policy advice must be tailored to the context where it is to be applied, and offer a 'Contextual Interaction Theory' to aid in analysis of what advice might be useful where and when. The consequences of the transfer of policy concepts to unsuitable contexts are reported in the case studies in Chapters 4 through 14. Enormous expenditures of resources and political capital are often involved in policy transfers. Yet, often, donors and sponsors, who probably exaggerated likely benefits of policy transfers, have little to show. Whether it is transferring concepts and techniques for dealing with flood risks to Romania, public participation in irrigation and hydropower in Turkey, or multi-stakeholder partnerships in Palestine, to name only a few such cases of such policy transfers, designs undergo considerable transformation as they are implemented, and many result in negligible change.

Following the guidance of the authors collected here will be challenging. An exceptionally strong network of water resources experts shares faith in hydro technology and for centuries has looked to structural solutions including dikes, dams, diversions and treatment plants to solve water problems. In the last few decades the water resources community increasingly emphasizes the role of governance as the greatest impediment to progress in solving water problems, and embraces a variety of policy advice including user-pays principles, cost of service pricing, privatization of utilities, decentralization and public participation (Conca, 2006). However, insights from the study of public policy and politics are only sometimes acknowledged by the water resources network of experts and seldom lead to changes in discourse and practice. As I have remarked elsewhere, 'the contemporary language of water resources reform tends to be rather bloodless and politically tone deaf' (Ingram, 2011, p.248). Policy changes most likely to be proposed have narrow, technical appeal that fails to motivate or engage elected leaders or broad social movements.

Adopting a contextual approach to policy transfer in water management requires a number of discomforting changes from past practice that are suggested in the conclusion to this book. Rather than expecting immediate results as measured by the improved state of water resources according to some objective rubric, the focus of evaluation must shift to evolution in water resources over the longer term with intermediate steps of changing narratives, mind-sets, and values and enhancing resources, capacity and power in particular settings. Instead of wholesale implementation of professionally crafted 'preferred' solutions, piecemeal adoption of some ideas, blended with local practice, will probably be necessary. Instead of insisting on conceptual clarity, where policy tools embodied in reforms appeal primarily to rational motivations and scientific logic, water resource experts may need to accommodate a wide variety of apparently inconsistent logics that bring into play diverse cultures and sense of place. Top-down delivery of policy reform must give way to transfer chains of actors in which each link is empowered to modify and, in consequence, 'own' the design. It is no surprise that successful reforms have many mothers and fathers, not so much because everyone wants to take credit, but because the best processes of policy transfer involve many people in various arenas.

As the authors in this volume argue, different kinds of analysts must become more closely involved in the design and implementation of contextually appropriate policy transfers. Among the social scientists in the existing professional community, law and economics have had the advantage in fashioning contemporary global water policy recommendations. Specialists in public policy, public administration, and political science have a small voice. When these disciplines are involved, it is most often as critics of past actions. In contextual analysis, these specialists have a more active role, engaging in close study of existing motives and values, networks of relationships, administrative structures and distribution of resources and power. Contextual Interaction Theory, as laid out in this book, insists that these kinds of factors receive as much scrutiny as does hydrology and other physical characteristics of basins and catchments. Only when context is taken into account in the design of policy reforms can policy transfer take place.

References

Conca, K. (2006) *Governing Water: Contentious Transnational Politics and Global Institutions Building*, The MIT Press, Cambridge, MA/London.

Ingram, H. (2011) 'Beyond universal remedies for good water governance: A political and contextual approach', in A. Garrido and H. Ingram (eds) *Water for Food in a Changing World*, Routledge, Abingdon.

Wildavsky, A. (1977) 'Doing better and feeling worse: the political pathology of health care', *Daedalus*, Winter, pp. 105–123.

Preface

Within this edited volume on contextual water management, we build on the work of a wide range of researchers who are each grappling with finding ways to address a plethora of water related issues and opportunities. Through our research, we as academics strive to develop an improved understanding of the world around us for numerous reasons – yet what we have in common is our desire to share our findings. This collection of case studies provides insight into the messy business of water management and enables the readers to draw a number of valuable lessons from them. Given that there is so much time and money being spent globally on transferring knowledge and best practices for water management, we and our team of researchers set out our task as being critical yet constructive about the processes we have been observing as part of our extensive work.

The editors and many of the authors are connected to CSTM, a policy studies research centre at the University of Twente, where we have a core group of researchers committed to understanding the contextual nature of water management. The contributors come from a wide variety of places and share a diversity of experiences and opinions. We think that both our common ground and diverse experiences are what makes this book so rewarding and valuable to the reader. No two contexts are alike, and understanding this from a fundamental perspective allows us to critically view the processes that we observe in terms of their applicability under different circumstances.

We, as editors, would like to thank our very dedicated and patient authors for travelling this road with us. By working together over a period of more than one year, we learned to respect and appreciate the different types of work that is done even within our own offices! Over the course of many revision rounds, authors worked with sincere determination to be both efficient and thorough in their responses to our 'detailed comments'. We are ever so grateful for your patience and truly excellent contributions! We would like to express our deepest appreciation to Helen Ingram for agreeing to write the foreword for this book. Additionally, we would like to thank Ada Krooshoop for her invaluable support during the process of delivering this manuscript.

For three of the four editors, this is our first attempt at editing a collection of chapters into a published manuscript. We found the experience to be extremely rewarding and significantly more work than we had expected. We do feel that our keen interest in providing a coherent, well-written and researched book has been satisfied through the provision of the work before you. This collection is intended to be of interest to other academics struggling with issues related to the transfer of policy and learning from the experiences of others. We hope that you enjoy reading our book!

List of abbreviations

ADB	Asian Development Bank
APFM	Associated Programme on Flood Management
AWM	Adaptive Water Management
BWAs	Basin Water Authorities
CAs	conservation authorities
CAM	Catchment Abstraction Management plan
CCPG	Guadalupe River Basin Commission
CIT	Contextual Interaction Theory
CONAGUA	National Water Commission Mexico
COP	Conference of the Parties
CRED	Centre for Research on the Epidemiology of Disasters
CSTM	Centre for Studies in Technology and Sustainable Development, University of Twente
CWR	Committee for Water Resources
DSI	Directorate General of State Hydraulic Works (*Devlet Su İşleri*)
EC	European Commission
ECHO	European Community Humanitarian Office
EEA	European Environmental Agency
EFRO	European Fund for Regional Development
EIA	Environmental Impact Assessment
EIU	Economist Intelligence Unit
EPA	Environmental Protection Agency
EU	European Union
EWASH	Emergency Water, Sanitation and Hygiene Group
EWB	Engineers without Borders
FAO	Food and Agriculture Organization
FLIWAS	Flood Information and Warning System
GAP	South-Eastern Anatolian Project
GoR	Government of Romania
GWP	Global Water Partnership
GWP-TAC	Global Water Partnership – Technical Advisory Committee

GSO	General Statistics Office of Vietnam
ICWE	International Conference on Water and the Environment
IMTA	Mexican Institute of Water Technology
IWRM	integrated water resources management
MBC	Mexican River Basin Council
Min. VROM	Ministry of Housing, Spatial Planning and the Environment
Min. V&W	(Dutch) Ministry of Transport, Public Works and Water Management
MSPs	multi-stakeholder partnerships
NAS	National Adaptation Strategy
NGOs	non-governmental organizations
NWP	Netherlands Water Platform
OECD	Organisation for Economic Co-operation and Development
OGWA	Ontario Ground Water Association
PCCC	Platform Communication on Climate Change
PET	polyethylene terephthalate
PHG	Palestinian Hydrology Group
PIANC	World Association for Waterborne Transport Infrastructure
PPPs	Public–Private Partnerships
PSDN	Peace, Security and Development Network
PSO/PSOM	Social Transformation Programme Matra and the Emerging Markets Cooperation Programme
PWA	Palestinian Water Authority
RBC	river basin councils
RMOW	Regional Municipality of Waterloo
RNE and EVD	Royal Netherlands Embassy and Agency for International Business and Cooperation
SAC	Special Areas of Conservation
SAGE procedure	Schémas d'Aménagement et de Gestion des Eaux – in English: Integrated Water Management Plan at River Basin Level
SCI	Sites of Community Importance
SDWA	Safe Drinking Water Act
SPA	Special Protection Areas
SPCs	Source Protection Committees
SPO	State Planning Office
SWP	source water protection
UK DfID	United Kingdom's Department for International Development
UN	United Nations
UNAM	National Autonomous University of Mexico

UNCED	United Nations Conference on Environment and Development
UNCSD	United Nations Conference on Sustainable Development
UNDP	United Nations Development Programme
UNDP-IWRM	United Nations Development Programme Project on Integrated Water Resources Management
UNFCCC	United Nations Framework Convention on Climate Change
UNPF	United Nations Populations Fund
UNECE	United Nations Economic Commission for Europe
UNEP	United Nations Programmes for Environment
UvW	Union of Water Boards
WASH	water supply, sanitation and hygiene
WASH+	water supply, sanitation and hygiene plus waste management
WB	World Bank
WFD	Water Framework Directive
WHO	World Health Organization
WMO	World Meteorological Organization
WUOs	water user organizations
WWAP	World Water Assessment Programme
WWF	World Wide Fund for Nature

1 Introduction

Gül Özerol, Cheryl de Boer and Joanne Vinke-de Kruijf

1.1 The international dimension of modern water management

Water is an important natural resource that has affected both the development and the destruction of societies since ancient times (Wittfogel, 1957; Postel, 1999). Modern societies are undergoing change on a number of fronts and thus the associated skills, knowledge and approaches for managing this vital resource are also changing. As such, water management is becoming an increasingly challenging endeavour.

The increasing difficulties associated with water management have arisen due to a variety of factors that can be grouped under four themes. First, a growing number of people are experiencing water quality and quantity problems, and will probably suffer from more severe water scarcity and pollution in the coming decades (WWAP, 2012). The impacts of diffuse uses and pollution are unlike the issues addressed in the past that could be dealt with solely by end-of-the-pipe solutions. The interactions of pollutants and other changes in the quality and quantity of water available are also increasingly complex and difficult to understand. Second, the frequency and severity of extreme events such as droughts and floods continue to increase due to climate change (Solomon *et al.*, 2007). As a result, water users, managers and policymakers have to deal with higher risks that are associated with these events. Third, domestic and industrial sectors are consuming higher levels of water due to the continuing global population growth and the ever-accelerating pace of urbanization (UN, 2011, 2012). However, irrigated agriculture remains the largest user sector in many arid and semi-arid countries, which also rely on irrigation for food security and improved livelihoods (Molden, 2007). Last but not least, the privatization of water sources and services, and the competition for water by various sectors bring up questions related to effective and equitable allocation (Swyngedouw, 2005).

Many countries thus face similar pressures on their water system. The international character and urgency of similar water problems imply a potential as well as a need to learn from each other's experiences. At the same time, the context in which water management problems arise largely differs between localities. For example, freshwater available per capita is less than 1,000 m^3 in Northern Africa and more than 70,000 m^3 in Canada (WWAP, 2012). Thus, there is a great potential and need for developing solutions at an international scale and for transferring water policies and knowledge from one locality to another. At the same time, there is also a necessity to look for context-specific solutions.

One of the major factors that complicate the transfer of modern water management knowledge is that water problems are often not primarily technical problems but rather governance problems. Water governance refers to the idea that water is managed through the interactions between many interdependent actors, both public and private. These actors operate at various levels, have different perceptions and objectives and employ various strategies and instruments. Hence, a characteristic of modern water governance systems is that they involve multiple elements, including levels, scales, actors, networks, problem perceptions, objectives, strategies, instruments, responsibilities and resources (Kuks, 2004). Modern water governance systems also acknowledge the growing number and connectedness of water uses and users and the growing uncertainties of the system as a result of, for example, climate change (Van der Brugge and Van Raak, 2007; Pahl-Wostl *et al.*, 2008). They therefore embrace the inherent complexity of water issues as opposed to fighting them (De Boer and Bressers, 2011), incorporate interdisciplinarity (Castro, 2007), include public participation and take the whole water system into account (Burua and Van Ast, 2011). While these ideas about modern water management are valuable, they are unlikely – especially when made operational – to be equally applicable in diverse natural and social settings. Thus, the transfer of knowledge should be done with care and only after a careful consideration of the context-specific conditions and circumstances in which the knowledge was developed (e.g. Mossberger and Wolman, 2003; Ingram, 2011; Swainson and de Loe, 2011).

While many scholars stress that knowledge developed in one context does not – without modification – apply to another context, there is a tendency, in particular among international and supranational institutions and organizations, to look to 'panaceas', i.e. universal remedies that promise to advance the general state of water management. These panaceas take many different forms such as blueprints, best practices, success stories, models and narratives (Meinzen-Dick, 2007; Molden, 2007; Molle, 2008; Mukhtarov, 2008; Ingram, 2011). They are often associated with modern water management concepts, such as integrated water resources management (including the establishment of water authorities at the river basin or watershed level), 'good' water governance and participatory water management. Further, they are strongly promoted – or even imposed – by organizations such as the European Commission, the World Bank and the Global Water Partnership. They are also spread within the scope of development aid or the cross-border work of research organizations, consultancies and governmental agencies. Despite their ambitious goal of addressing the relevant water management problems, the extent to which the incorporation and implementation of these panaceas have been able to achieve this goal in practice needs to be debated. Two major features of panaceas give rise to this need. First, they treat management alternatives as mutually exclusive, implying that if an alternative is chosen, the others are often completely left out (Meinzen-Dick, 2007). This limited view of managing water leads to failures that sometimes take decades to appear (Mollinga, 2008). Second, the added value of panaceas is overestimated over the other possible options. Panacea approaches assume rigid definitions and boundaries of what is 'good' and what is 'bad'for water management. Narratives are often used as discursive tools to make these definitions and draw boundaries for disseminating panaceas (Molle, 2008).

Thus, despite the widespread efforts to develop and transfer knowledge about modern water management and governance, the analysis and assessment of the real value of these efforts is an important area for further research. This observation has led the editors and contributors of this book to present not only a wide variety of international case studies that involve a transfer of water management knowledge, but also to examine with an objective and critical eye the emergence and application of key concepts in modern water management such as integrated water resources management, river basin management, good water governance, participatory water management and adaptive water management.

1.2 Policy transfer and contextual water management

Although no current evidence was found that policy transfers are on the rise, we believe this to be quite likely in the context of water management since the number of and influence arising from international organizations are increasing alongside a rapid growth in communication means (Dolowitz and Marsh, 2000). While there has been limited attention paid to transfers among water management scholars, this phenomenon has been studied intensively in political sciences and international studies. Policy transfer is a generic concept that can be defined as a process by which policy-relevant knowledge developed in one context is used in another (Evans, 2004; Dolowitz and Marsh, 2000). The term 'policy-relevant knowledge' refers to knowledge associated with the solving of public problems and includes knowledge about policies, institutions and ideas as well as concrete expertise or programmes (Stone, 1999). Context refers to the setting of a transfer, or the conditions and circumstances that influence a transfer process. In the case of water management, this includes the political and institutional contexts, the cultural and economic contexts as well as the biophysical and geographical contexts (Swainson and de Loe, 2011). In this book, we use the term 'policy transfer' to describe not only transfer processes that involve 'action-oriented intentional learning' (Evans, 2009) but also less intentional or unintentional transfer processes that concern, for example, the translation of global discourses into national policies or the emergence of best practices.

The concept of 'policy transfer' is closely related to and partly overlaps with the concepts of 'technology transfer' and 'knowledge transfer'. Policy transfers imply by definition knowledge transfers and *can* take the form of a technology transfer. The reason for using the term 'policy transfer' rather than 'technology transfer' or knowledge transfer' is that the former is commonly used to describe transfers in the public domain whereas the latter terms are more common in organization sciences. Transfers can occur across space (between several political systems) or time (within the same political system) (Dolowitz and Marsh, 2000; Rose, 1993), across sectors or government levels (Evans, 2004) and also – in the case of technology or knowledge transfer – across employees of the same organization or different organizations (Reddy and Zhao, 1990; Trott *et al.*, 1995). This book focuses on international transfers within the same sector, namely the water management sector, that occur between actors representing different countries and/or governance levels. This includes transfers in which actors of one country transfer knowledge to/from actors of another country as well as transfers in which national or subnational actors translate knowledge to/from the supranational level. As we believe that actors, their activities and interactions are of crucial importance in policy transfers, we conceptualize policy transfer as an interactive process among the actors. Regarding the objects being transferred in a water management context, we focus on knowledge about water management and water governance. This knowledge can take the form of policies, programmes, ideas, concepts, methods, technologies and the like. Thus, in this book, policy transfer refers to *the interactive process by which actors use knowledge (about water management and/or governance) that was developed in one context (country or governance level) for the development of water management and/or water governance in another context.* In analysing these transfers, this book uses the perspective of 'context matters' as a starting point, which explains the reasoning behind the choice of 'International studies on contextual water management' as the subtitle of the book. The main assertion underlying the notion of 'contextual water management' is that actors who shape the course and outcomes of a transfer process are influenced by the context in which they act (Bressers and Lulofs, 2010; Bressers, 2004; Bressers, 2009; De Boer and Bressers, 2011). Among the central aims of this book is therefore to understand how, where and why context affects actors involved in a transfer process and what this effect implies for the way that the relevant knowledge should be received under these circumstances.

1.3 Objectives and approach of this book

Recognizing the importance of contextual water management, this book aims to critically discuss the transferability of modern water management and governance concepts by analysing the contextual factors specific to the particular transfer being studied. The academic merit of such an endeavour is that it increases the precision and understanding of what is known about policy transfers and how policy transfers can best be placed in the appropriate context for study and learning. The practical relevance of such an analysis is that it helps to understand the specific context under which modern water expertise is derived and thus enables practitioners to better determine the value and the context under which it is best exported. Through this book, we hope to make policy designers, consultants and researchers more aware of the pitfalls of unprepared (or even well informed and careful) export and dissemination of ideas.

This book builds upon a collection of eleven international studies. To provide a breadth of examples from which different contextual insights can be garnered, studies are chosen to show a wide range of different national contexts and transfers. Some studies are based on multiple case studies whereas others address only single cases. All of the chapters include reflections on the transferability of knowledge about water management or governance. This does not mean that all of the cases describe or are based on an actual transfer. Some studies are rather examples of prospective evaluations, implying that the transferability of an object is assessed before the transfer takes place (Mossberger and Wolman, 2003). Some studies also question whether the process under study actually reflects a transfer or not. Various chapters further critically reflect on how a transfer benefits its receivers and pay special attention to explicit discourses and narratives that are used by the actors involved.

This book is not based on a comparative research design to avoid imposing a simplistic framework to the heterogeneous studies. We do however seek to compare the practical examples that are presented in various chapters along a number of lines of interest. To enable a valuable comparison while not forcing a strict pre-definition of the types of transfers included, all authors were asked to reflect on the following questions, when applicable:

1 What characterizes the transfer process? What idea is transferred and how, if at all? Who or what promotes the idea? To what extent is the transfer voluntary or coercive?
2 What characterizes the context(s) of the study? In which context was the object of transfer developed? To what context was or could the object of transfer be transferred? What are the differences between these contexts?
3 What is the effectiveness or added value of the transfer? What makes the transfer (potentially) successful or beneficial, or not? For what reasons is the transfer successful or beneficial, or not?

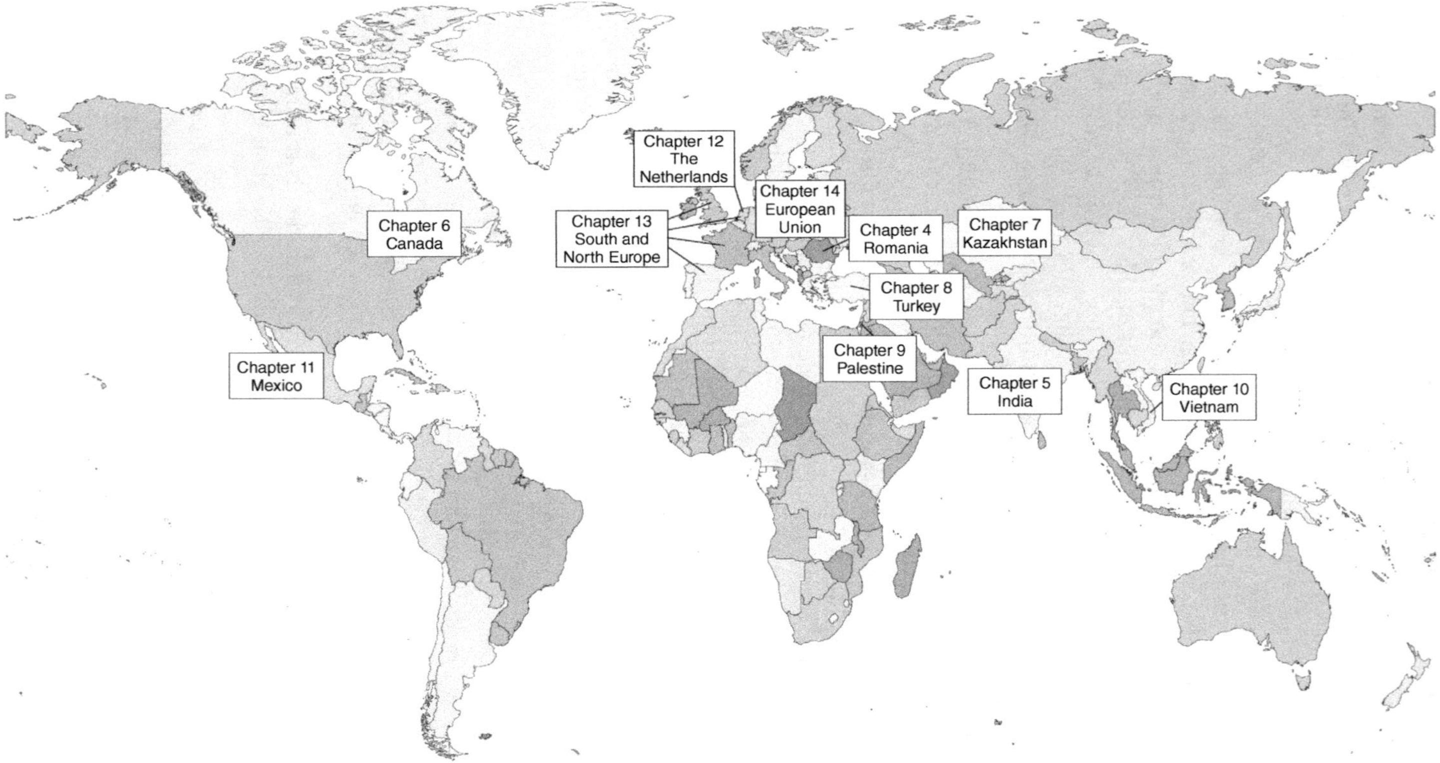

Figure 1.1 Geographical location of the cases included in the book.

Our case study selection did not aim to include specific continents, regions or countries. Nevertheless, the book has a wide geographic coverage with cases from three continents (Europe, North America and Asia). Figure 1.1 shows the geographic location of the cases that are included in various studies. Moreover, the book elaborates on transfers regarding a variety of water management issues such as flood risk management, drought, urban agriculture, drinking water provision and climate change adaptation. Incorporating a diverse range of transfers related to water governance and management is valuable for two reasons. First, it reflects the diversity occurring in the field of policy transfer. Second, it reflects the complexity and interconnectedness between the interactions taking place before, during and after policy transfers. Since the literature available regarding this more critical view of policy transfer is limited, we envision this breadth as serving the purpose of enlightening further studies into different aspects and types of transfer.

Throughout the chapters of this book, a number of different yet prevalent ideas can be found related to modern water management and governance. Additionally, the chapters incorporate context-specific elaborations of the variations that are experienced and understood related to the theoretical ideas about policy and knowledge transfer. As such, the theoretical underpinnings about contextual water management and governance that have been developed at CSTM (Twente Centre for Studies in Technology and Sustainable Development at the University of Twente, the Netherlands) are supported in the following chapters with a breadth of theories, frameworks and ideas.

1.4 Structure of the book

Given the plethora of theoretical concepts that are applicable to understanding the interactions taking place around the world in the name of improved water management, the editors have chosen to include two theoretical chapters to collect this vast expanse of literature and explore the various ways that improved water management is given meaning. Chapter 2, the first theoretical chapter, introduces three key concepts in modern water management – integrated water resources management, good water governance and participatory water management – and explains the concept of policy transfer. Following this, Chapter 3 provides an introduction of the Contextual Interaction Theory. A number of authors use the theory as the conceptual lens through which they view the interactions taking place in their cases. This theory enables a parsimonious analysis of these interaction processes and facilitates the inclusion of a wide variety of contextual factors to be systematically taken into the analysis.

The theoretical chapters are followed by eleven empirical chapters that offer a multitude of interesting theory-guided case studies. For easy reference throughout the book, Table 1.1 provides an overview of the contents of these chapters.

Table 1.1 Overview of the empirical chapters

Chapter	*Location*	*Case study*	*Type of knowledge*	*Scope of chapter*
4	Romania	Three regional projects supported by the Netherlands	Concepts, methods and technologies in flood risk management	Influence of context on the effectiveness of a transfer process
5	India	Local project supported by the United States	Water drainage method (soak pit)	Assessment of the collaborative effort
6	Canada	Provincial level, prospective evaluation	Tools, techniques and procedures in the protection of drinking water sources	Contextual considerations shaping transferability
7	Kazakhstan	National level, translation of global discourses	Water policy innovations including IWRM and river basin organizations	Role of international and local discourses
8	Turkey	National and local levels, contestation of a global discourse	Public participation in irrigation management and hydropower planning	Contested nature of a concept and influence of context
9	Palestine	Local projects, implemented with international donors	Multistakeholder partnership approach in water supply and waste management	Influence of fragile setting on the applicability and effectiveness of transfer
10	Vietnam	Provincial level, transfer of international commitments	Integration of agriculture and wastewater management in urban areas	Role of local governance systems on the lack of transfer
11	Mexico	Basin level, transfer of international management principles	River basin councils (basin management and public participation)	Influence of contextual factors on the efficacy of transfer
12	The Netherlands	National and local levels, translation of an emerging discourse	Adaptation to climate change in a water management context	Influence of local context on adoption process
13	Four European countries	Basin level, change of governance regimes	Adaptive responses to drought and water deficiency	Role of contextual factors in achieving an adaptive regime
14	European member states	(national levels, prospective evaluation)	Integration of nature policies and water management (Natura 2000 and Building with Nature)	Transferability of emerging best practice

As Table 1.1 demonstrates, the collection of empirical chapters analyses the phenomenon of 'policy transfer' from different perspectives. Some chapters examine the transfer of policy knowledge that is established through methods, technologies, expertise, know-how, etc. Other chapters reflect upon the transfer of international concepts or emerging concepts. The chapters are organized accordingly into three different parts:

1 transferring established knowledge (Chapters 4 to 6)
2 transferring an international concept (Chapters 7 to 11)
3 transferring an emerging concept (Chapters 12 to 14)

In Chapter 15, the editors conclude the book with a reassessment of the value and validity of policy transfer in the global water management sector and provide a preliminary assessment of the importance of contextual factors observed throughout the chapters.

References

Bressers, H. and Lulofs, K. (2010) 'Analysis of boundary judgments in complex interaction processes', in H. Bressers and K. Lulofs (eds) *Governance and Complexity in Water Management: Creating Cooperation through Boundary Spanning Strategies*, Edward Elgar Publishing, Cheltenham, UK/Northampton, MA.

Bressers, H.T.A. (2004) 'Implementing sustainable development: How to know what works, where, when and how', in W.M. Lafferty (ed.) *Governance for Sustainable Development: The Challenge of Adapting Form to Function*, Edward Elgar Publishing, Cheltenham.

Bressers, H.T.A. (2009) 'From public administration to policy networks: Contextual interaction analysis', in S. Nahrath and F. Varone (eds) *Rediscovering Public Law and Public Administration in Comparative Policy Analysis: Tribute to Peter Knoepfel*, Presses Polytechniques et Universitaires Romandes, Lausanne.

Burua, S. and Van Ast, J.A. (2011) 'Towards interactive flood management in Dhaka, Bangladesh', *Water Policy*, vol. 13, no. 5, pp. 693–716.

Castro, J.E. (2007) 'Water governance in the twentieth-first century', *Ambiente & Sociedade*, vol. 10, no. 2, pp. 97–118.

De Boer, C. and Bressers, H. (2011) *Complex and Dynamic Implementation Processes: The Renaturalization of the Dutch Regge River*, University of Twente with the Dutch Water Governance Centre, Enschede.

Dolowitz, D.P. and Marsh, D. (2000) 'Learning from abroad: The role of policy transfer in contemporary policy making', *Governance*, vol. 13, no. 1, pp. 5–23.

Evans, M. (2004) 'Understanding policy transfer', in M. Evans (ed.) *Policy Transfer in Global Perspective*, Ashgate Publishing Limited, Aldershot.

Evans, M. (2009) 'Policy transfer in critical perspective', *Policy Studies*, vol. 30, no. 3, pp. 243–268.

Ingram, H. (2011) 'Beyond universal remedies for good water governance: A political and contextual approach', in A. Garrido and H. Ingram (eds) *Water for Food in a Changing World*, Routledge, Abingdon.

Kuks, S.M.M. (2004) 'Water governance and institutional change', PhD thesis, University of Twente at Enschede, The Netherlands.

Meinzen-Dick, R. (2007) 'Beyond panaceas in water institutions', *Proceedings of the National Academy of Sciences*, vol. 104, no. 39, pp. 15200–15205.

Molden, D. (ed.) (2007) *Water for Food, Water for Life: A Comprehensive Assessment of Water Management in Agriculture*, Earthscan with the International Water Management Institute, London.

Molle, F. (2008) 'Nirvana concepts, narratives and policy models: Insights from the water sector', *Water Alternatives*, vol. 1, no. 1, pp. 131–156.

Mollinga, P. (2008) 'Water policy–water politics: social engineering and strategic action in water sector reform', in W. Scheumann, S. Neubert and M. Kipping (eds) *Water Politics and Development Cooperation: Local Power Plays and Global Governance*, Springer Verlag, Berlin and Heidelberg.

Mossberger, K. and Wolman, H. (2003) 'Policy transfer as a form of prospective policy evaluation: Challenges and recommendations', *Public Administration Review*, vol. 63, no. 4, pp. 428–440.

Mukhtarov, F. (2008) 'Intellectual history and current status of integrated water resources management: A global perspective', in C. Pahl-Wostl, P. Kabat and J. Möltgen (eds) *Adaptive and Integrated Water Management: Coping with Complexity and Uncertainty*, Springer Verlag, Berlin and Heidelberg.

Pahl-Wostl, C., Tàbara, D., Bouwen, R., Craps, M., Dewulf, A., Mostert, E., Ridder, D. and Taillieu, T. (2008) 'The importance of social learning and culture for sustainable water management', *Ecological Economics*, vol. 64, no. 3, pp. 484–495.

Postel, S. (1999) *Pillar of Sand: Can the Irrigation Miracle Last?* WW Norton & Company, New York and London.

Reddy, N.M. and Zhao, L. (1990) 'International technology transfer: A review', *Research Policy*, vol. 19, no. 4, pp. 285–307.

Rose, R. (1993) *Lesson-Drawing in Public Policy: A Guide to Learning across Space and Time*, Chatham House Publishers, Chatham, NJ.

Solomon, S., Qin, D., Manning, M., Chen, Z., Marquis, M., Averyt, K.B., Tignor, M. and Miller, H.L. (eds) (2007) *Climate Change 2007: Working Group I: The Physical Science Basis. Contribution of Working Group I to the Fourth Assessment Report of the Intergovernmental Panel on Climate Change*, Cambridge University Press, Cambridge and New York.

Stone, D. (1999) 'Learning lessons and transferring policy across time, space and disciplines', *Politics*, vol. 19, no. 1, pp. 51–59.

Swainson, R. and de Loe, R.C. (2011) 'The importance of context in relation to policy transfer: A case study of environmental water allocation in Australia', *Environmental Policy and Governance*, vol. 21, no. 1, pp. 58–69.

Swyngedouw, E. (2005) 'Dispossessing H_2O: The contested terrain of water privatization', *Capitalism Nature Socialism*, vol. 16, no. 1, pp. 81–98.

Trott, P., Cordey-Hayes, M. and Seaton, R.A.F. (1995) 'Inward technology transfer as an interactive process', *Technovation*, vol. 15, no. 1, pp. 25–43.

UN (2011) 'World Population Prospects – The 2010 Revision', Department of Economic and Social Affairs, United Nations, New York.

UN (2012) 'World Urbanization Prospects – The 2011 Revision', Department of Economic and Social Affairs, United Nations, New York.

Van der Brugge, R. and Van Raak, R. (2007) 'Facing the adaptive management challenge: insights from transition management', *Ecology and Society*, vol. 12, no. 2, art. 33.

Wittfogel, K.A. (1957) *Oriental Despotism: A Comparative Study of Total Power*, Yale University Press, New Haven, CT.

WWAP (World Water Assessment Programme) (2012) '*The United Nations World Water Development Report 4: Managing Water under Uncertainty and Risk*', UNESCO, Paris.

2 Water management solutions

On panaceas and policy transfer

Joanne Vinke-de Kruijf and Gül Özerol

2.1 Introduction

With the increase of communication means, it has become easier than ever for policymakers, water managers, researchers and other actors involved in water management to collaborate. As explained in Chapter 1, this collaborative environment creates opportunities for developing best practices and transferring water management solutions from one context to another. This chapter elaborates on some of the 'universal remedies' and on the concept of policy transfer in a water management context. First, we discuss three concepts that are often promoted as 'the solution' to water management problems: integrated water resources management, good water governance and participatory water management. In addition to reviewing the common definitions and understandings of these concepts, we also reflect on the issue of putting them in practice. This elaboration of the key concepts in modern water management is followed by a discussion about the phenomenon of 'policy transfer'. After discussing the phenomenon itself and explaining what, why and who transfers, we pay specific attention to the outcomes of a transfer process and the role of contextual factors. The chapter closes with some concluding remarks that reflect on how this chapter relates to Chapter 3, which forms together with this chapter the theoretical basis of the empirical chapters of the book. The discussions presented in this chapter are based on a literature review of scholarly articles as well as international reports and policy documents. Where relevant, the review extends beyond water management to the broader literature on policy studies and environmental decision-making. The ideas introduced in this chapter are critically reflected upon in Chapter 15, which is the concluding chapter of this book.

2.2 Key concepts in water management

'Integration' – integrated water resources management

The need to integrate multiple elements into water management has been acknowledged since 1950s (Biswas, 2008; White, 1998). The Dublin

Conference in 1992, however, gave the impetus to integrate the ecological, social and economic pillars of sustainable development in water management (Rahaman and Varis, 2005). At this conference, the so-called 'Dublin Principles', which aim to integrate the ecological, social and economic pillars of sustainable development into water management, were accepted (ICWE, 1992).

The concept of integrated water resources management (IWRM) has been increasingly promoted, mainly by the Global Water Partnership (GWP), which was established in 1996 to support all the interested actors in sustainable water management. According to the common and often-cited definition of the GWP, IWRM is '*a process which promotes the coordinated development and management of water, land and related resources, in order to maximize the resultant economic and social welfare in an equitable manner without compromising the sustainability of vital ecosystems*' (GWP-TAC, 2000, p. 22).

In the last two decades, attempts to put IWRM into practice have increased with an impetus from international and supranational organizations (Anderson *et al.*, 2008). IWRM gained wide acceptance from international organizations such as the World Water Council, the World Bank, the Organisation for Economic Co-operation and Development (OECD), the United Nations Programmes for Environment and Development (UNEP and UNDP), mainly due to the fact that it builds on the Dublin Principles. At the European Union (EU) level, the Water Framework Directive (WFD) was enacted to unite both the disaggregated water-related legislation and to respond to the concerns about degrading water quality in the EU region (Kallis and Nijkamp, 2000). Social and economic integration are also among the major objectives of the WFD as reflected with the inclusion of provisions about public participation, the polluter-pays principle and full-cost water pricing (European Parliament and Council, 2000).

River basins in IWRM are both literally and conceptually seen as the 'natural' unit of management. This situation makes river basin management a crucial element of IWRM. The EU and the GWP have been the major promoters of river basin management. The WFD has specific provisions to realize the river basins as the management units, whereas the GWP deems river basins as the 'logical planning units for IWRM from a natural system perspective' (GWP-TAC, 2000, p. 24). Using the hydrological boundaries as management units is as a manifestation of the hydrology-based approach of IWRM. This hydrological orientation is well accepted by some scholars (Jaspers, 2003; Mostert *et al.*, 2007), whereas there is also criticism and caution. The major argument is that, as a hydrological concept, the river basin contradicts the integration ideal by putting hydrology as the main discipline of water management and by undervaluing the role of other disciplines (Allan, 2003; Merrey, 2005).

Despite the worldwide attempts to put IWRM into practice, there are arguments that IWRM cannot be fully implemented since integration is impossible due to its high ambitions (Biswas, 2008; Grigg, 2008). Including other resources in the scope of integration, let alone the inevitable complexity caused by the multitude of policy sectors, scales, uses, users, disciplines and professions, remains a challenge. Biswas (2008) argues that neither the definition nor the operationalization of IWRM incorporates the view of scientific disciplines related to land resources. However, every land use decision is also a water use decision (Bossio *et al.*, 2010). This implies that IWRM cannot be realized without integrating policy decisions about the use of land resources. Furthermore, the inherent complexity of social-ecological systems requires the involvement of policymakers, scientists and practitioners for the integration of knowledge and expertise about water and land resources (Duda and El-Ashry, 2000).

Another point of caution about IWRM is its 'political correctness'. Political issues in water management include the asymmetries in the distribution of power, the access of different actors to water resources and the distribution of benefits as well as costs among the multiple stakeholders. However, these issues are often ignored while integrating the multiple interests and values (Molle, 2008; Mollinga, 2008; Ingram, 2011). Ignoring the inherent political nature of water management processes and professional discourses could make IWRM a 'sanctioned discourse' that is imposed by some hegemonic actors onto other less powerful ones (Allan, 2003).

Regarding the political dimension of river basin management, Moss and Newig (2010) warn that rescaling the management level from administrative units to river basins creates winners and losers by reshaping the power relations. The winners and losers of the rescaling process are not only the actors of water management, but also the other policy sectors, since the misfit between the jurisdictional boundaries and hydrological boundaries can lead to problems in other sectors such as environmental protection (Cohen and Davidson, 2011; Ingram, 2011). Integration of multiple policy sectors is not unique to river basin management, rather an overarching problem for IWRM. In many countries, sectoral approaches dominate the whole policymaking realm. Therefore, supporting the necessary institutional arrangements and ensuring the allocation of resources and authority at multiple levels are identified as relevant issues to tackle (Falkenmark *et al.*, 2007; Imperial, 2009).

The pessimistic views about IWRM relate to the difficulty in overcoming the obstacles required to go from IWRM theory to practice (Jeffrey and Gearey, 2006; Biswas, 2008). Nevertheless, all the criticisms concerning IWRM point out the need to learn from the practical realities of IWRM implementation to create flexible and enabling management approaches (Coenen and Bressers, 2012).

'Good governance' – good water governance

The concept of good governance is derived from 'governance', which closely resembles 'government', but should be understood as a separate concept. In public management debates, governance has been increasingly articulated to replace 'government', yet still with diverse definitions and uses. With a caution that governance is no more a synonym to government, Rhodes (1996) identifies six uses of the concept: the minimal state; corporate governance; good governance; the new public management; socio-cybernetic systems; and self-organizing networks. In most of these uses, governance implies the increased involvement of non-governmental actors in decision-making processes and the extension of the stakeholder range along the institutional and jurisdictional levels. The concept of good governance is particularly relevant, since it has been applied in all policy sectors, including water resources.

Good governance is promoted largely by the World Bank, which uses the concept and its associated principles as the criteria to improve the effectiveness of development aid in government reforms. According to the World Bank good governance is

> a predictable, open, and enlightened policymaking (that is, transparent processes); a bureaucracy imbued with a professional ethos; an executive arm of government accountable for its actions, and a strong civil society participating in public affairs; and all behaving under the rule of law.
>
> (World Bank, 1994)

The World Bank identifies four interrelated components of assessing governance, namely public sector management; accountability; legal framework for development; and transparency and information (World Bank, 1994). These components form the basis of good governance and have been used by different organizations. Other international organizations such as the OECD, the International Monetary Fund and UNDP also embrace the concept. Furthermore, various new topics such as corruption, public participation and human rights, have been incorporated into development cooperation (Hoebink, 2006). The definition and the components of good governance reflect the normative stance behind the concept through defining how governance should look like. The concept of 'governance' alone does not have such a connotation.

Similar to IWRM, the GWP also promotes the concept of water governance. In their formative report, Rogers and Hall (2003, p. 7) cite the GWP's water governance definition as '*the range of political, social, economic and administrative systems that are in place to develop and manage water resources, and the delivery of water services, at different levels of society*'. This definition is crucial as it clearly addresses how water governance relates to water management. On the one hand, water management is concerned with developing and utilizing the water resources as well as delivering the water services. On the other hand, water governance sets the rules of operation for water management (Rogers and Hall, 2003; WWAP, 2006).

Good water governance combines the notions of good governance and water governance. However, the concept of good water governance also has many definitions with no consensus about its meaning and use (Lautze *et al.*, 2011). One of the fundamental assumptions is that good water governance brings about good outcomes, which reflects the normative element of good governance. What constitutes good water governance and good outcomes, however, remains unclear.

Expectations concerning the application of good governance principles to the water domain vary. These expectations include the establishment of a common understanding about the 'good outcomes' of governance processes and the collaboration of different sectors towards achieving those good outcomes (Rogers and Hall, 2003; Molden *et al.*, 2010). As governance is about the exercise of authority and allocation of rights and resources, the issue of social power becomes difficult to avoid. However, water governance is also subject to criticism when it is considered only as an instrument for achieving IWRM and the application of good water governance remains apolitical by neglecting how the decisions are made by respecting – or disrespecting – the principles (Mollinga, 2008; Lautze *et al.*, 2011).

'Public participation' – participatory water management

The participation of lay public members and organized stakeholders is crucial for sustainable development. This perspective is reflected in several international policy documents. In the Rio Declaration, the benefit expected from public participation is expressed in terms of its contribution to dealing with sustainability issues (UNCED, 1992). Agenda 21, which was prepared as the main outcome of the Rio Declaration, addresses the necessity of public participation at all levels for successful implementation. The need for participation in policy formulation, decision-making and implementation is stated in the Johannesburg Declaration as well (UN, 2002). Accepted in 1998, the Aarhus Convention reflects the international consensus for improving participatory mechanisms in the general realm of environmental decision-making (UNECE, 1998). The Aarhus Convention grants citizens and organizations the right to be informed about environmental matters and to participate in environmental decision-making. Thus, an increasing emphasis has been put on designing and implementing participatory mechanisms in the decision-making processes towards sustainable development.

The emphasis put on public participation can be explained by a multitude of expected benefits for participants and governments. Two main arguments, namely normative and instrumental, expound upon these expected benefits (Coenen, 2009). The normative argument asserts that participation is necessary for better functioning democracies, since it serves normative ideals such as equity, representativeness, transparency and empowerment (Fiorino, 1990; Webler, 1995; Leach, 2006). The instrumental argument, on the other hand, stresses the role of participation as a tool to increase the legitimacy of policy decisions and to improve the effectiveness of policy implementation (Beierle and Cayford, 2002; Coenen, 2009). In line with the instrumental view, a third argument is referred to as the substantive argument (Fiorino, 1990). This argument suggests that the judgements of lay public members are as sound as expert knowledge and therefore should be incorporated to improve the quality of decisions (Beierle and Cayford, 2002).

Participatory mechanisms that incorporate the values, beliefs and knowledge of the public are also regarded as an indispensable element of decision-making processes in water management. This perspective is shared by water scholars (Jaspers, 2003; Delli Priscoli, 2004; Creighton, 2005; Von Korff *et al.*, 2012) as well as the international organizations that promote IWRM (GWP-TAC, 2000) and good water governance (WWAP, 2012).

On the international stage, public participation found wider acclaim in water management starting in the 1990s. At the Dublin Conference, participation was identified as one of the four guiding principles for water management (ICWE, 1992). In line with the Dublin Principles, the GWP, a key promoter of IWRM, denotes public participation as a key principle of IWRM (GWP-TAC, 2000). Being the major legislation of the EU water policy, the WFD also requires the implementation of various participatory mechanisms in all phases of water management (EC, 2002).

Along with the ambitious expected benefits of public participation, several concerns lead to questions about achieving these benefits in water management (Mostert, 2003; Özerol and Newig, 2008; Neef, 2009). For instance, when participants lack competency or knowledge, their involvement can prove futile in making the necessary decisions, cause confusion for the decision makers, or lengthen the decision-making process. Another concern is the utilization of the resources allocated to a participation process. From an instrumental perspective, the use of time, money and human resources would be inefficient if a successful outcome is not reached despite the implementation of participatory mechanisms (Özerol and Newig, 2008; Krywkow, 2009). While public participation can be an appropriate means of empowerment and improved democracy, it can also turn into a bureaucratic exercise to fulfil procedural requirements or a marketing tool to 'sell' government policy (Mostert, 2003). Participation can also become a managerial exercise with which the procedures or techniques from elsewhere are applied without reflecting on the role of different types of actors (Cleaver, 1999).

The obstacles against the achievement of the expected benefits of public participation can be attributed to numerous factors. These include poor planning of participation processes, complexity and vagueness of higher level participation mandates for the local actors, inexperience about participatory mechanisms, differing perceptions about the expected outcomes and reluctance in putting participation into practice (Steelman and Ascher, 1997; Cooke and Kothari, 2001).

Thus, participation has – like integration and good governance – many potential benefits but these benefits are difficult to achieve in practice. As various empirical chapters of this book show, the implementation of participatory mechanisms does not necessarily lead to the improved management and governance of water resources.

2.3 Towards an interpretation of policy transfer processes

Policy transfer and related concepts

In recent years, an extensive body of literature has been formed around the analysis of processes that involve the 'borrowing of ideas'. As explained in Chapter 1, we use the term 'policy transfer' in this book to point towards the interactive process by which actors use knowledge that was developed in another context. The term policy transfer encompasses a range of related concepts. In some of the literature, the phenomenon is studied from the perspective of policymakers searching for new ideas across borders. These voluntary transfers are also referred to as 'lesson-drawing' (Rose, 1993) or 'systemically pinching ideas' (Schneider and Ingram, 1988). Later publications on policy transfer (Dolowitz and Marsh, 1996, 2000; Stone, 1999) or 'institutional transplantation' (De Jong *et al.*, 2002a) also include less voluntary forms of policy transfer. In this book, the term policy transfer also encompasses the 'translation' of ideas or concepts from the supranational or international level to the national, regional or local level. In its original form, the concept of 'translation' refers to how the spread of anything (e.g. claims, goods or artefacts) – in space and time – is influenced by people (Latour, 1986). In this book, the concept refers to the international spreading of ideas and is associated with 'global discourses' (Fadeeva, 2004; Mukhtarov, 2012). The concept of 'policy transfer' is also closely related to 'policy diffusion' (Walker, 1969). Research on policy diffusion, however, has another focus. It concentrates on the structural factors that explain the adoption of an innovation and is often based on large-*n* quantitative studies. Policy transfer research, on the other hand, concentrates on process-oriented questions including the how, when and why of diffusion and tends to be based on in-depth case studies (Dolowitz and Marsh, 1996; Mossberger and Wolman, 2003). Two other concepts that are closely related to policy transfer are 'policy convergence' (Bennett, 1991) and 'institutional isomorphism' (DiMaggio and Powell, 1983). Convergence and isomorphism both refer to the idea that institutions or policies have the tendency to become alike. Policy transfer is a potential cause of this phenomenon. However, convergence may also occur unintentionally, for example, due to harmonizing macroeconomic forces. Policy transfer, on the other hand, always involves an agent (Evans, 2009).

Transfers are also studied extensively in the domain of organizational sciences. In this domain, the phenomenon is usually referred to as 'knowledge transfer' or 'technology transfer'. Transfer studies in this domain focus on the transfer of knowledge (about technologies or the like) across branches of international firms (Bresman *et al.*, 1999; Reddy and Zhao, 1990), from developed to developing countries, for instance, in a developmental aid setting (Pigram, 2001; Siggel, 1986; Svensson, 2007) or across organizations in relation to innovation (Carlile, 2004; Trott *et al.*, 1995). These studies have proven to provide useful insights for the analysis of transfers in the public domain. For example, the concept of institutional isomorphism was developed by organizational theorists (DiMaggio and Powell, 1983) and later also used to understand policy transfer in the European Union (Radaelli, 2000). Another example is the four stages of a technology transfer (i.e. awareness, association, assimilation and application), which were developed to understand technology transfers between firms (Trott *et al.*, 1995) and recently also applied to the diffusion of innovations in the public sector (Bressers, 2011). Moreover, literature on knowledge management and knowledge sharing has proven to provide a better understanding of how various types of knowledge and interaction mechanisms influence the transfer of knowledge in a water management context (Vinke-de Kruijf *et al.*, 2011). What knowledge is being transferred in a water management context, why and by whom is elaborated further in the next subsections.

Knowledge being transferred

Central in policy transfers are the transfer of policy-relevant *knowledge.* The reason for using the term 'knowledge' is to emphasize that it is not the concepts, methods or technologies themselves that are transferred but rather the knowledge that is associated with or embedded in these objects. Consider, for example, the transfer of a technology from one country to another. On such occasions, it is not the equipment itself but the knowledge that is associated with that equipment that is transferred from one context to another (Bresman *et al.*, 1999; Trott *et al.*, 1995). Objects of policy transfer include: (1) policy aspects (e.g. goals, content or instruments); (2) concrete programmes; (3) institutions; and (4) ideologies, ideas and attitudes (Dolowitz and Marsh, 2000; Rose, 1993; Stone, 1999). Policy aspects include both 'broader statements of intention' regarding the direction policymakers want to take and instruments or administrative techniques used to achieve a certain policy. Programmes are mentioned separately as they are concrete courses of action used to implement a policy (Dolowitz and Marsh, 1996, 2000). The transfer of policies or programmes is not limited to administrative aspects; it also includes the transfer of technical knowledge, for example, in the form of design aspects (Marsden and Stead, 2011). Transfers may further concern institutions, which are the patterns that structure social interaction. These patterns consist of formal rules specifying what is allowed, obliged and forbidden under what conditions and informal social practices and rituals that are rooted in cultural norms and values. A policymaker may also transfer an ideological rhetoric to justify the adoption or spread of certain policies (Dolowitz and Marsh, 1996). Think tanks can enhance such transfers by legitimizing or underpinning policies with intellectual matter, rhetoric or a scholarly discourse (Stone, 2000). In addition, negative lessons are often mentioned as a separate category (Dolowitz and Marsh, 1996, 2000) to emphasize that lessons can be positive and negative. Policies can be a stimulus for some countries to adopt similar policies and at the same time become a negative example for other countries (Rose, 1993).

The transferability of knowledge is likely to depend on the type of knowledge being transferred. For example, methods and techniques are more transferable than less visible knowledge (e.g. ideas) or more context-specific knowledge (e.g. programmes) (Stead, 2012). Thus, informal practices are more difficult to transfer than formal procedures. The transferability of knowledge also relates to different levels of action at which they occur. These levels of action include the constitutional level (ground rules), the level of policy areas (relations between governmental bodies) and the operational level (daily activities). Transfers are likely to be more demanding for higher levels of action than for lower levels. Accordingly, legal systems are more difficult to transfer than practical procedures (De Jong and Mamadouh, 2002). Transfers occur not only between actors at similar levels but also across levels. Actors at higher levels can take lessons from actors at lower levels, while actors at lower levels can learn from actors at higher levels too (Dolowitz and Marsh, 2000). In theory, policy transfer may take place at and between any of the following levels: transnational, international, national, regional and local. Accordingly, policy transfers may occur along twenty-five different pathways. Of these pathways, the transfers that involve an international dimension are the most interesting ones as they cannot be explained with other frameworks. Hence, this book and much of the relevant literature focus on transfers that involve an international or transnational dimension (Evans and Davies, 1999).

Triggers and causes behind transfer processes

Policy transfers are often understood as processes consisting of several phases. While various scholars identify a different number of phases, they broadly agree that policy transfers start with an exploratory phase (this phase may include the mobilization and emergence of an actor network), which is followed by an assessment phase and potentially results in an application phase (Bressers, 2011; Evans and Davies, 1999; Jeffrey and Seaton, 2004; Mossberger and Wolman, 2003; Rose, 1993; Trott *et al.*, 1995). According to these scholars, a search for knowledge is central in the exploratory phase. However, transfers are not necessarily rooted in a desire to acquire or import knowledge. They can also be rooted in a desire to spread or export knowledge (Stone, 1999). From an institutional transplantation perspective, the main reason behind any type of transfer is the expectation of bringing improvements to the receiving country. They are seen as ways to speed up developments or to achieve them at lower costs (Mamadouh *et al.*, 2002). Other scholars argue that transfers start with the recognition of a problem (Evans and Davies, 1999) or dissatisfaction with the status quo (Rose, 1993). In recent years, there has been critique on this interpretation of policy transfer. One of these critiques is that transfers are seen as rational policy processes in which policy decisions are the realization of valued goals through structured interventions (James and Lodge, 2003). Another critique is that policy transfers are hardly ever take the form of free and rational responses to emerging problems. Actors face bounded rationality since they do not have the access or the capacity to process all information (Dolowitz and Marsh, 2000). Besides that actors have incomplete knowledge, they are also bounded by organizational choices (James and Lodge, 2003). Policy transfers are also not necessarily based on deliberate choices. They are sometimes also rooted in political choices. Actors may, for example, just introduce 'foreign' ideas to get things on the political agenda or as a 'quick fix' in response to political pressure or a crisis. On such occasions, policy transfers are initiated even though actors have no idea how well the idea will work or what adaptations will be needed (Mossberger and Wolman, 2003).

One of the critiques on 'lesson-drawing' is furthermore that it fails to acknowledge that transfers often involve coercion, i.e. they involve situations in which one actor forces or pushes another actor to adopt a certain policy. Although the direct imposition of policies is rare, transnational institutions, such as the World Bank or the International Monetary Fund, are able to force governments to adopt certain policies. In addition, indirect coercive transfers are quite common. For example, interdependencies or the perception of lagging behind may force or push governments to adopt similar policies (Dolowitz and Marsh, 1996). Consider, for example, the desire of Eastern European countries to catch up with Western countries after communism (Rose, 1993). In reality, policy transfers therefore often involve both voluntary and coercive elements. They are neither based on perfect rationality nor on direct imposition but include aspects of bounded rationality, a perceived necessity or conditionality (Dolowitz and Marsh, 2000).

In voluntary transfers, the power balance is generally more on the side of the receiver. In such cases, transfers can take the form of highly selective borrowing (often from multiple sources), undiluted borrowing (little adaptation to local circumstances) or synthetic innovation (policies are synthesized with other policies to create more advanced innovations). In coercive or imposed transfer, the power balance is more at the side of the source. In such cases, policy transfers take the form of imposition. Such imposed forms of policy transfers are typical in a colonial or in a repressive communist context. However, authoritarian imposition without any consideration of local conditions does not automatically occur in such contexts. As imposition does not prevent a transfer from being disagreed upon, it is likely to result in contested or negotiated forms of imposition (Ward, 1999).

Policy transfers are often closely related to processes of policy convergence or institutional isomorphism. Policy convergence is based on the recognition that countries face similar problems and tend to solve them in similar ways (Bennett, 1991). One possible cause behind policy convergence is penetration, which is similar to direct coercive transfer and involves the compulsion to conform or the use of power. A second possible cause is emulation or policy bandwagoning, which refers to the borrowing and subsequent adaptation of policies. A third cause is harmonization, which refers to the recognition of interdependence and is promoted and sustained by supranational institutions, such as the European Union. Fourth, policy convergence can be caused by the networking and interaction among 'elite' (e.g. experts or practitioners) (Bennett, 1991; Stone, 2000). Another factor that explains policy transfers is institutional isomorphism, which refers to the idea that organizations have the tendency to become alike. This especially plays a role in transnational structures, such as the European Union. Isomorphism can take the form of coercion and also of mimetism or normative pressure. Mimetism refers to imitation as a way of coping with uncertainty. Normative pressure arises from the fact that professionals tend to have a common cognitive base and a shared legitimization. In a European context, coercion is not likely to dominate but it plays a role via economic mechanisms. Given the level of uncertainty, mimetic isomorphism is necessary in many circumstances. Normative pressure tends to occur in policy processes that are dominated by expert knowledge (Radaelli, 2000). Thus, policy transfers occur for many different reasons and are triggered by different factors. The types of explanations that are provided closely relate to the adopted theoretical lens for studying the phenomenon.

The actors involved

One of the aspects that distinguishes the literature on – both voluntary and coercive – policy transfers from the other related literature streams, such as policy diffusion or policy convergence, is that it sees transfers as intentional activities. Thus, as processes that occur through the actions of certain agents. These agents may work at different levels (i.e. the international, transnational, national, regional or local level) and at different points in the process (Evans and Davies, 1999). To understand the type of actors involved in policy transfer, we distinguish four categories of actors: (1) supranational and international actors; (2) governmental actors; (3) non-governmental actors; and (4) private actors. These categories are not mutually exclusive and are partly overlapping. For example, the last three categories are likely to include both national and international actors (cf. Stone, 2004).

The first category of actors involves representatives of supranational or international organizations (Dolowitz and Marsh, 1996, 2000). The terms international and supranational refer here to structures and processes that involve two or more countries. International actors play an important role in the diffusion of ideas, both informally and formally (Mossberger and Wolman, 2003). This category therefore includes actors with formal power, such as the EU or the UN, and actors with informal power, such as the GWP, the World Water Council or the Stockholm International Water Institute. It also includes funding agencies, such as the International Monetary Fund and regional and international development banks. Other examples of actors in this category are international non-governmental organizations such as the World Wildlife Fund and Greenpeace or international social movements like Occupy. The importance of these non-state transnational organizations in policymaking is argued to be on the increase (Evans and Davies, 1999).

The category of governmental actors includes elected officials, political parties and civil servants (Dolowitz and Marsh, 1996, 2000). Civil servants (bureaucrats or non-elected officials) can play a significant role in policy transfers, as they are usually part of informal communities of experts that share ideas, information and values through publications, conferences or organizations. The membership of these communities often cuts across boundaries of organizations, sectors and countries (Rose, 1993).

Further, policy transfers often engage non-governmental or semi-public organizations. These actors represent non-profit organizations and usually have no role in formal decision-making or in administration or government policies (Stone, 2000). Examples of actors in this category are think tanks, research institutes, universities, pressure groups and voluntary organizations (Stone, 1999, 2000) as well as lobby or interest groups (Marsden and Stead, 2011). These actors may either be focused more on research or engaged in advocacy and marketing of ideas. What they have in common is that they have a significant degree of autonomy and are intellectually independent. Because of their information, involvement, extensive network and expertise, they can play an important role in the diffusion of ideas. However, the actual influence of non-state actors, such as think tanks, on policy transfer is often limited. They can construct legitimacy, develop knowledge, assess policy options and draw attention but are dependent on formal political actors for policy transfers to occur (Stone, 2000). This does not apply to policy transfer networks (i.e. collaborative structures of state and non-state actors). These networks are often key instruments in policy learning and can fulfil an important function in knowledge dissemination (Evans, 2009).

Private actors also often contribute to the transfer of policies. Policymakers often depend on the advice of consultants who act as experts in the development of new policies or programmes (Dolowitz and Marsh, 2000). What also plays a role is that society is increasingly shaped by the interaction between various actors, both public and private (Bressers and Kuks, 2003; Peters and Pierre, 1998). In particular in the management of urban water resources, the private sector becomes increasingly important due to the privatization of water services (Swyngedouw, 2005).

Transfer studies are often based on the idea that policy transfers basically involve two actors: a source that is transferring knowledge to a receiver. In reality, policy transfers occur in a much more networked setting (De Jong and Edelenbos, 2007) and are realized through the interaction between various actors (Evans and Davies, 1999). One of the underlying reasons is that the solving of public problems usually requires a combination of various resources owned by different actors. Problems are therefore dealt with by mutually dependent actors that have their own, often diverging, interests, problem perceptions and strategies (Koppenjan and Klijn, 2004). Policy transfers are therefore more likely to be shaped by multiple actors who either fulfil the role of source (donor or exporting actor), receiver (host or importing actor) (De Jong and Mamadouh, 2002; Vinke-de Kruijf *et al.*, 2011), facilitator (bringing various actors together) or producer (supporting the transfer process by developing knowledge). Every actor can fulfil multiple roles (Wolman and Page, 2002) at various points in time (Ward, 1999).

2.4 The outcomes and context of policy transfers

Assessment of a policy transfer

Among the central questions that guide the empirical chapters of this book is a question concerning the success of the transfer under study (see Chapter 1). This question becomes of particular relevance when 'policy transfer' is seen not only as a lens to study the travel of ideas (as is often done in diffusion studies) but also as an intentional activity. Depending on what one defines as the purpose of a policy transfer, one may adopt different evaluation criteria to assess a transfer. Marsden and Stead (2011) argue that there are very few studies that demonstrate an actual transfer. In their opinion, this would involve an evaluation of the extent to which a search for external knowledge resulted in new policies or changes to proposed policies. From a policy implementation perspective, the demonstration of an actual transfer lies in the actual application of knowledge and thus in the implementation phase. In analysing the actual outcomes of a policy transfer, a distinction can be made between four gradations of policy transfer: copying, emulation, combination and inspiration. Copying involves an actor adopting an object (i.e. a policy, programme or institution) without modifying anything. Such transfers are rather rare but may occur if, for example, a politician is looking for a 'quick fix'. In an international setting, copying is nearly impossible as differences in language and legal procedures can hardly be ignored. Emulation means that an object is seen as the 'best standard'. It does not involve the transfer of the object itself but the transfer of the ideas behind an object. Combination (or hybridization) is one of the most typical forms of policy transfer. It means that actors combine elements found in various settings or objects to develop an object that fits the local context. It thus involves adjustment for contextual differences as well as combining elements from different places. Inspiration means that foreign objects inspire fresh thinking and policy change. However, the final outcome does not draw directly upon the original object (Dolowitz and Marsh, 2000; Evans, 2009; Rose, 1993). In various phases of a transfer, obstacles may arise preventing the actual transfer of knowledge. Problems may already arise in the pre-decision phase. For example, actors may fail to define or recognize a problem or are not receptive to alternatives. Also during the transfer process, actors may experience obstacles such as a lack of support, structural constraints or technical problems (Evans, 2009).

The evaluation of policy transfers in terms of policy implementation can be problematic. The ultimate outcomes of a transfer process (i.e. concrete changes in the environmental or societal context) often only become visible over the long term. Hence, it is of added value to also analyse the degree to which a process changed the characteristics of actors involved (Vinke-de Kruijf *et al.*, 2012). Such changes are reflected, for example, in conceptual replication. Transfers are interactive processes in which knowledge, ideas and information are transferred from one individual to another. Conceptual replication is indicated through the increasing number of known concepts and the adaptation of existing concepts, which can be used eventually in the policy process (De Jong and Edelenbos, 2007). Similar evaluation criteria are central in the literature in which policy transfers are seen as learning processes that result in an increase in policy-relevant knowledge. In analysing the effects of learning, a distinction can be made between: (1) governmental actors learning about processes resulting in organizational change; (2) policy networks learning about instruments resulting in programme change; and (3) policy communities learning about ideas resulting in paradigm shifts (Bennett and Howlett, 1992). Real policy learning involves knowledge also being taken into account in the policy process. This implies that, for example, organizational learning can only occur if the transfer process involves actors who are in the position to shape organizational behaviour (Wolman and Page, 2002). The realization of an actual transfer is not necessarily an indication of success. An exploratory phase may also teach actors what not to do (negative lessons) or bring actors to the conclusion that there are – at least at that moment – too many obstacles (Rose, 1993).

In general, caution is needed when assuming a causal relation between policy learning and policy change (e.g. modification of policies or adoption of new policies). Policy change may result from many factors other than policy learning and policy learning may not be expressed in and not even be intended for policy change. If policy transfer processes are perceived as learning processes, effects can also become visible in other factors, such as the acquisition of new factual knowledge, changes in norms, values and belief systems or in enhanced trust and improved understanding of other actors' mindsets (Huitema *et al.*, 2010). To avoid the problems associated with the evaluation of policy change, one could choose to limit an evaluation to the process itself, for example, by evaluating 'collaboration' (as is done in Chapter 5). Another alternative approach is to evaluate transfers in actor-specific terms. Such an evaluation questions, for example, the degree to which the transfer contributed to the realization of the initial objectives of the receiving or the transferring actor or the degree to which the transfer is perceived as a success (Dolowitz and Marsh, 2000).

On the role of contextual factors

As explained in Chapter 1, this book concentrates on policy transfers with an international dimension. Such transfers often involve actors from different contexts. This characteristic of international policy transfers has received little attention in policy studies. It is, however, important to recognize that the transfer of knowledge tends to already be problematic between actors of the same country and to increase with geographic and cultural distance (Bresman *et al.*, 1999). The opportunity for knowledge transfer is higher in cases where actors have a more common background in terms of socio-cultural inheritance, organizational belonging and profession (Stenmark, 2002). Because of their diverging backgrounds, actors involved in policy transfers often have difficulties in arriving at mutual understanding through communication. This may not only exert a negative influence on the process itself but also contribute to the failure of intended policy transfers (Vinke-de Kruijf *et al.*, 2011, 2012).

Another characteristic of international transfers is that they involve the transfer of knowledge from one context to another. Compared to transfers within the same country, such transfers are rather challenging as differences in political systems and policy environment are likely to be more pronounced (Mossberger and Wolman, 2003). What also plays a role is that countries differ greatly in terms of resources and socio-economic development, political values and also in terms of their awareness of foreign nations and cultures (Rose, 1993). Thus, given the differences in both natural and social contexts, it is rather unlikely that concepts such as IWRM are equally valid and applicable for different countries (Biswas, 2004). Scholars therefore widely acknowledge that policy transfer should be done with care and only after a careful consideration of the context-specific conditions and circumstances in which it was developed (e.g. Ingram, 2011; Mossberger and Wolman, 2003; Swainson and de Loe, 2011).

When focusing on structural constraints, one may argue that transfers tend to be easier between countries that are more similar in political, legal and cultural terms. This assumption seems plausible but has not been confirmed in comparative studies. In fact, transfers between dissimilar countries often appear to be much easier than transfers between countries that belong to the same 'family of nations' (De Jong *et al.*, 2002a). The underlying reason is that transfers between similar countries are subject to the same pitfalls as transfers between dissimilar countries. In spite of common legal, cultural and political characteristics they often also face different practical and institutional issues. Therefore even transfers between similar contexts should consider and adapt policy-relevant knowledge to national, regional and local factors and needs (De Jong, 2004). Thus, transfers require the modification of knowledge and involve learning processes (Stead, 2012). From a policy translation perspective, modification is not only necessary but also an inevitable component of policy change. It emphasizes that any type of policy change involves translation, i.e. the modification of meaning in the travel of knowledge (Mukhtarov, 2012).

The contextual nature of policy was confirmed in various studies. For example, comparative research shows that participatory approaches in water management are much more effective in some countries than in others. Context-specific institutional factors, such as a lack of experience with multi-actor approaches, fear of losing control, the distribution of water rights or controversies between authorities, partly explain this (Mostert *et al.*, 2007). Research on water governance reforms in the former Soviet Union sketch a similar picture. While reforms were implemented, legacies of old structures (such as a hierarchical culture, strong fragmentation and a lack of horizontal coordination) remained. Actors only implemented those reforms that were socially appropriate or economically attractive while neglecting reforms that were incompatible (e.g. more public participation) (Sehring, 2009). Thus, transfers are influenced by the rather stable cultural, institutional and historical configurations of a country. Historical developments determine what is currently possible and existing political cultures cannot be abandoned easily (Rose, 1993). Similar arguments are used, for example, to explain why policy transfers from Western to Eastern Europe often failed. The underlying problem was that changes in institutional 'hardware' (i.e. the formal structure of rules, rights procedures and principles) were not supported by institutional 'software' (i.e. prevailing discourses) (Dryzek, 1996).

On the role of the specific context and actors

There has been little systemic research on the actual effectiveness of policy transfers. On the basis of their experiences, Dolowitz and Marsh (2000) observe three broad reasons for policy failure: (1) insufficient information about the object (uninformed transfer); (2) non-transfer of crucial elements (incomplete transfer); and (3) insufficient attention to contextual differences (inappropriate transfer) (Dolowitz and Marsh, 2000). A study in which policy transfer is analysed as a form of prospective evaluation highlights similar aspects. The study is based on the recognition that the quality of a transfer crucially depends on the possibility to assess the potential applicability of the object in another setting. Such an assessment depends first and foremost on the quality of information. Is there accurate information about the policy to be transferred and about the relative success of that policy at various locations? It further involves that the actors involved reflect on the similarities and differences between different settings. To what extent is the policy addressing similar problems and goals and is it applied in a similar setting? (Mossberger and Wolman, 2003).

As explained in the previous section, literature highlights that the success of a transfer depends in particular on whether knowledge is modified to fit local conditions. One would therefore expect that voluntary transfers are more effective than coercive transfers. Comparative research does not confirm this expectation. What is relatively more important is that powerful actors support a transfer (De Jong *et al.*, 2002b). The relative importance of individual actors was confirmed in other studies. Transfers between dissimilar settings do not have to fail and transfers between similar settings are more effective when they engage powerful receivers who are convinced that the transfer is useful and have a strong desire to change things. Transfers tend to fail if local actors are not in favour, are hardly involved, have no opportunity to adjust the model or if the transfer is imposed (i.e. the model is only accepted because of the funds associated with engaging in the process) (Kroesen *et al.*, 2007). Furthermore, flexible approaches that consider loosely defined or multiple models are more successful than approaches that involve the copying of a model. The latter approaches tend to invoke local resistance and/or low levels of compliance (De Jong *et al.*, 2002b). Along similar lines, De Jong (2004) found that the effectiveness of a transfer depends on whether actors consider and – if necessary – synthesize multiple models and are able to create a sense of urgency and strong coalition that pushes the initiative. Thus, the actors involved and the specific context of a transfer exert a strong influence on the effectiveness of a policy transfer.

2.5 Concluding remarks

Policy transfer can take many forms and manifests itself in many ways. In a water management context, transfers are manifested not only in the transfer of knowledge from one country to another but also in the spreading of 'universal remedies'. Earlier in this chapter, we have discussed some of these remedies in water management. We have also discussed how these concepts are promoted by international organizations, the meanings that are attached to them and the difficulty of putting them into practice. Relying on any policy concept as a panacea can prove to be a hopeless effort towards solving the water management problems, since panaceas are like the 'optimum solutions' that are impossible to reach with multiple and conflicting objectives. Maybe, as Ingram (2011) suggests, we need 'clumsy solutions' that embrace multiple types of knowledge and appeal to different values. In search for such solutions, sharing of knowledge and experiences can be particularly promising but should be done with care. The concept of policy transfer and related concepts have been reflected on in the previous sections to better understand the benefits and limitations of transfers. We show that they may concern different types of knowledge, result from different factors and involve various actors. We also touched upon the evaluation of policy transfer and the role of structural and specific factors. What becomes clear is that context matters but that contextual differences or constraints in itself do not explain the success or failure of a transfer – assuming that one can demonstrate such 'success'. Policy transfer is unlikely to be successful without some form of translation, i.e. modification or adaptation of knowledge to fit the context in which knowledge is being applied.

Understanding the differences and similarities of context is thus essential when studying policy transfers. It is also essential to understand how the specific policy intervention will and can be absorbed into the receiving context. For such an understanding, the interaction processes between the different levels and scales of governance as well as the implementation processes that will be associated with the final stage of the policy transfer need to be explored. The next chapter provides a basis for understanding these interaction processes. The Contextual Interaction Theory is elaborated as a way of understanding the important characteristics of actors and how they influence an interactive process, including the planning and implementation of a transfer. In discussing contextual factors attention is paid not only to the wider context of such processes but also to the governance context in which these transfers occur. The importance of the qualities of the governance regime for the receiving context are also highlighted, which provides some opportunities to assess the likelihood of the path of implementation solely based on the receiving end, as no matter how similar or different the contexts are, some are just more receptive to incoming ideas and changes than are others.

References

Allan, T. (2003) 'IWRM/IWRAM: A new sanctioned discourse?' Occasional Paper 50, School of Oriental and African Studies, London, pp. 1–27.

Anderson, A, Karar, E. and Farolfi, S. (2008) 'Synthesis: IWRM lessons for implementation', *Water SA*, vol. 34, no. 6, pp. 665–669.

Beierle, T.C. and Cayford, J. (2002) *Democracy in Practice: Public Participation in Environmental Decisions*, Resources for the Future, Washington, DC.

Bennett, C.J. (1991) 'What is policy convergence and what causes It?', *British Journal of Political Science*, vol. 21, no. 2, pp. 215–233.

Bennett, C.J. and Howlett, M. (1992) 'The lessons of learning: Reconciling theories of policy learning and policy change', *Policy Sciences*, vol. 25, no. 3, pp. 275–294.

Biswas, A.K. (2004) 'Integrated water resources management: A reassessment', *Water International*, vol. 29, no. 2, pp. 248–256.

Biswas, A.K. (2008) 'Integrated water resources management: Is it working?' *Water Resources Development*, vol. 24, no. 1, pp. 5–22.

Bossio, D., Geheb, K. and Critchley, W. (2010) 'Managing water by managing land: Addressing land degradation to improve water productivity and rural livelihoods', *Agricultural Water Management*, vol. 97, no. 4, pp. 536–542.

Bresman, H., Birkinshaw, J. and Nobel, R. (1999) 'Knowledge transfer in international acquisitions', *Journal of International Business Studies*, vol. 30, no. 3, pp. 439–462.

Bressers, H.T.A. and Kuks, S.M.M. (2003) 'What does "governance" mean? From conception to elaboration', in H. Bressers and W.A. Rosenbaum (eds) *Achieving Sustainable Development: The Challenge of Governance across Social Scales*, Praeger Publishers, Westport, CT.

Bressers, N. (2011) 'Co-creating innovation: A systemic learning evaluation of knowledge and innovation programmes', PhD thesis, Erasmus University, Rotterdam.

Carlile, P.R. (2004) 'Transferring, translating, and transforming: An integrative framework for managing knowledge across boundaries', *Organization Science*, vol. 15, no. 5, pp. 555–568.

Cleaver, F. (1999) 'Paradoxes of participation: questioning participatory approaches to development', *Journal of International Development*, vol. 11, no. 4, pp. 597–612.

Coenen, F.H.J.M. (ed.) (2009) *Public Participation and Better Environmental Decisions*, Springer, Dordrecht.

Coenen, F.H.J.M. and Bressers, H.T.A. (2012) 'Trends, drivers and dilemmas in the transition towards sustainable water management', in J. Meadowcroft, O. Langhelle and A. Ruud (eds) *Governance, Democracy and Sustainable Development: Moving Beyond the Impasse*, Edward Elgar Publishing, Cheltenham/Northampton, MA.

Cohen, A. and Davidson, S. (2011) 'An examination of the watershed approach: Challenges, antecedents, and the transition from technical tool to governance unit', *Water Alternatives*, vol. 4, no. 1, pp. 1–14.

Cooke, B. and Kothari, U. (eds) (2001) *Participation: The New Tyranny?* Zed Books, New York.

Creighton, J.L. (2005) 'What water managers need to know about public participation: One US practitioner's perspective', *Water Policy*, vol. 7, no. 3, pp. 269–278.

De Jong, M. (2004) 'The pitfalls of family resemblance', *European Planning Studies*, vol. 12, no. 7, pp. 1055–1068.

De Jong, M. and Edelenbos, J. (2007) 'An insider's look into policy transfer in transnational expert networks', *European Planning Studies*, vol. 15, no. 5, pp. 687–706.

De Jong, M. and Mamadouh, V. (2002) 'Two contrasting perspectives on institutional transplantation', in M. de Jong, K. Lalenis and V. Mamadouh (eds) *The Theory and Practice of Institutional Transplantation: Experiences with the Transfer of Policy Institutions*, Kluwer Academic Publishers, Dordrecht.

De Jong, M., Lalenis, K. and Mamadouh, V. (eds) (2002a) *The Theory and Practice of Institutional Transplantation: Experiences with the Transfer of Policy Institutions*, Kluwer Academic Publishers, Dordrecht.

De Jong, M., Mamadouh, V. and Lalenis, K. (2002b) 'Drawing lessons about lesson drawing: What the case reports tell us about institutional transplantation', in M. de Jong, K. Lalenis and V. Mamadouh (eds) *The Theory and Practice of Institutional Transplantation: Experiences with the Transfer of Policy Institutions*, Kluwer Academic Publishers, Dordrecht.

Delli Priscoli, J. (2004) 'What is public participation in water resources management and why is it important?' *Water International*, vol. 29, no. 2, pp. 221–227.

DiMaggio, P.J. and Powell, W.W. (1983) 'The iron cage revisited: Institutional isomorphism and collective rationality in organizational fields', *American Sociological Review*, vol. 48, no. 2, pp. 147–160.

Dolowitz, D.P. and Marsh, D. (1996) 'Who learns what from whom: A review of the policy transfer literature', *Political Studies*, vol. 44, no. 2, pp. 343–357.

Dolowitz, D.P. and Marsh, D. (2000) 'Learning from abroad: The role of policy transfer in contemporary policy making', *Governance*, vol. 13, no. 1, pp. 5–23.

Dryzek, J.S. (1996) 'The informal logic of institutional design', in R.E. Goodin (ed.) *The Theory of Institutional Design*, Cambridge University Press, Cambridge.

Duda, A.M. and El-Ashry, M.T. (2000) 'Addressing the global water and environment crises through integrated approaches to the management of land, water and ecological resources', *Water International*, vol. 25, no. 1, pp. 115–126.

EC (European Commission) (2002) *Guidance on Public Participation in Relation to the Water Framework Directive. Active Involvement, Consultation, and Public Access to Information, Common Implementation Strategy Guidance Document No. 8*, European Commission, Luxembourg.

European Parliament and Council (2000) Water Framework Directive: Directive 2000/60/EC, *Official Journal of the European Communities*, Brussels.

Evans, M. (2009) 'Policy transfer in critical perspective', *Policy Studies*, vol. 30, no. 3, pp. 243–268.

Evans, M. and Davies, J. (1999) 'Understanding policy transfer: A multi-level, multi-disciplinary perspective', *Public Administration*, vol. 77, no. 2, pp. 361–385.

Fadeeva, Z. (2004) 'Translation of sustainability ideas in tourism networks: Some roles of cross-sectoral networks in change towards sustainable development', *Journal of Cleaner Production*, vol. 13, no. 2, pp. 175–189.

Falkenmark, M., Finlayson, C.M. and Gordon, L.J. (2007) 'Agriculture, water, and ecosystems: Avoiding the costs of going too far', in D. Molden (ed.) *Water for Food, Water for Life: A Comprehensive Assessment of Water Management in Agriculture*, Earthscan, London and International Water Management Institute, Colombo, pp. 233–277.

Fiorino, D.J. (1990) 'Citizen participation and environmental risk: A survey of institutional mechanisms', *Science, Technology, and Human Values*, vol. 15, no. 2, pp. 226–243.

Grigg, N.S. (2008) 'Integrated water resources management: Balancing views and improving practice', *Water International*, vol. 33, no. 3, pp. 279–292.

GWP-TAC (Global Water Partnership – Technical Advisory Committee) (2000) *Integrated Water Resources Management, TAC Background Papers No. 4*, Global Water Partnership, Stockholm.

Hoebink, P. (2006) 'European donors and "good governance": Condition or goal?' *The European Journal of Development Research*, vol. 18, no. 1, pp. 131–161.

Huitema, D., Cornelisse, C. and Ottow, B. (2010) 'Is the jury still out? Toward greater insight in policy learning in participatory decision processes – the case of Dutch citizens' juries on water management in the Rhine Basin', *Ecology and Society*, vol. 15, no. 1, art. 16.

ICWE (International Conference on Water and the Environment) (1992) *The Dublin Statement on Water and Sustainable Development*, Dublin.

Imperial, M.T. (2009) Paradoxes, possibilities, and the obstacles to integrated water resources management: Lessons from the institutional rational choice literature, International Symposium on Society and Resource Management, 5–8 July 2009, Vienna.

Ingram, H. (2011) 'Beyond universal remedies for good water governance: A political and contextual approach', in A. Garrido and H. Ingram (eds) *Water for Food in a Changing World*, Routledge, Abingdon.

James, O. and Lodge, M. (2003) 'The limitations of "policy transfer" and "lesson drawing" for public policy research', *Political Studies Review*, vol. 1, no. 2, pp. 179–193.

Jaspers, F.G.W. (2003) 'Institutional arrangements for integrated river basin management', *Water Policy*, vol. 5, no. 1, pp. 77–90.

Jeffrey, P. and Gearey, M. (2006) 'Integrated water resources management: Lost on the road from ambition to realisation?' *Water Science and Technology*, vol. 53, no. 1, pp. 1–8.

Jeffrey, P. and Seaton, R.A.F. (2004) 'A conceptual model of "receptivity" applied to the design and deployment of water policy mechanisms', *Environmental Sciences*, vol. 1, no. 3, pp. 277–300.

Kallis, G. and Nijkamp, P. (2000) 'Evolution of EU water policy: A critical assessment and a hopeful perspective', *Zeitschrift für Umweltpolitik and Umweltrecht*, vol. 23, no. 3, pp. 301–335.

Koppenjan, J.F.M. and Klijn, E. (2004) *Managing Uncertainties in Networks: A Network Approach to Problem Solving and Decision Making*, Routledge, London.

Kroesen, O., De Jong, M. and Waaub, J.P. (2007) 'Cross-national transfer of policy models to developing countries: Epilogue', *Knowledge, Technology and Policy*, vol. 19, no. 4, pp. 137–142.

Krywkow, J. (2009) 'A methodological framework for participatory processes in water resources management', PhD thesis, University of Twente, Enschede.

Latour, B. (1986) 'The powers of association', in J. Law (ed.) *Power, Action and Belief: A New Sociology of Knowledge*, Routledge, London.

Lautze, J., De Silva, S., Giordano, M. and Sanford, L. (2011) 'Putting the cart before the horse: Water governance and IWRM', *Natural Resources Forum*, vol. 35, no. 1, pp. 1–8.

Leach, W.D. (2006) 'Collaborative public management and democracy: Evidence from western watershed partnerships', *Public Administration Review*, vol. 66, no. 1, pp. 100–110.

Mamadouh, V., De Jong, M. and Lalenis, K. (2002) 'An introduction to institutional transplantation', in M. de Jong, K. Lalenis and V. Mamadouh (eds) *The Theory and Practice of Institutional Transplantation: Experiences with the Transfer of Policy Institutions*, Kluwer Academic Publishers, Dordrecht.

Marsden, G. and Stead, D. (2011) 'Policy transfer and learning in the field of transport: A review of concepts and evidence', *Transport Policy*, vol. 18, no. 3, pp. 492–500.

Merrey, D.J. (2005) 'Integrating "livelihoods" into integrated water resources management: Taking the integration paradigm to its logical next step for developing countries', *Regional Environmental Change*, vol. 5, no. 4, pp. 197–204.

Molden, D., Lautze, J., Shah, T., Bin, D., Giordano, M. and Sanford, L. (2010) 'Governing to grow enough food without enough water: Second best solutions show the way', *International Journal of Water Resources Development*, vol. 26, no. 2, pp. 249–263.

Molle, F. (2008) 'Nirvana concepts, narratives and policy models: Insights from the water sector', *Water Alternatives*, vol. 1, no. 1, pp. 131–156.

Mollinga, P.P. (2008) 'Water policy – water politics: Social engineering and strategic action in water sector reform', in W. Scheumann, S. Neubert and M. Kipping (eds) *Water Politics and Development Cooperation: Local Power Plays and Global Governance*, Springer, Berlin/Heidelberg, pp. 1–29.

Moss, T. and Newig, J. (2010) 'Multilevel water governance and problems of scale: setting the stage for a broader debate', *Environmental Management*, vol. 46, no. 1, pp. 1–6.

Mossberger, K. and Wolman, H. (2003) 'Policy transfer as a form of prospective policy evaluation: Challenges and recommendations', *Public Administration Review*, vol. 63, no. 4, pp. 428–440.

Mostert, E. (2003) 'The challenge of public participation', *Water Policy*, vol. 5, no. 2, pp. 179–197.

Mostert, E., Pahl-Wostl, C., Rees, Y., Searle, B., Tàbara, D. and Tippett, J. (2007) 'Social learning in European river-basin management: Barriers and fostering mechanisms from 10 river basins', *Ecology and Society*, vol. 12, no. 1, art. 19.

Mukhtarov, F. (2012) 'Rethinking the travel of policy ideas: Policy translation in the water sector', *Policy and Politics*, online, available as a pre-press fast track article at: www.ingentaconnect.com/content/tpp/pap/pre-prints/15799;jsessionid=1frvpse9zxz69.alice.

Neef, A. (2009) 'Transforming rural water governance: Towards deliberative and polycentric models?', *Water Alternatives*, vol. 2, no. 1, pp. 53–60.

Özerol, G. and Newig, J. (2008) 'Evaluating the success of public participation in water resources management: Five key constituents', *Water Policy*, vol. 10, no. 6, pp. 639–655.

Peters, B.G. and Pierre, J. (1998) 'Governance without government? Rethinking public administration', *Journal of Public Administration Research and Theory*, vol. 8, no. 2, pp. 223–243.

Pigram, J.J. (2001) 'Opportunities and constraints in the transfer of water technology and experience between countries and regions', *International Journal of Water Resources Development*, vol. 17, no. 4, pp. 563–579.

Radaelli, C.M. (2000) 'Policy transfer in the European Union: Institutional isomorphism as a source of legitimacy', *Governance*, vol. 13, no. 1, pp. 25–43.

Rahaman, M.M. and Varis, O. (2005) 'Integrated water resources management: Evolution, prospects and future challenges', *Sustainability: Science, Practice and Policy*, vol. 1, no. 1, pp. 15–21.

Reddy, N.M. and Zhao, L. (1990) 'International technology transfer: A review', *Research Policy*, vol. 19, no. 4, pp. 285–307.

Rhodes, R.A.W. (1996) 'The new governance: Governing without government', *Political Studies*, vol. 44, no. 4, pp. 652–667.

Rogers, P. and Hall, A.W. (2003) 'Effective water governance', Technical Committee Background Papers No.7, Global Water Partnership Technical Committee, Stockholm.

Rose, R. (1993) *Lesson-Drawing in Public Policy: A Guide to Learning Across Space and Time*, Chatham House Publishers, Chatham, NJ.

Schneider, A. and Ingram, H. (1988) 'Systematically pinching ideas: A comparative approach to policy design', *Journal of Public Policy*, vol. 8, no. 1, pp. 61–80.

Sehring, J. (2009) 'Path dependencies and institutional bricolage in post-Soviet water governance', *Water Alternatives*, vol. 2, no. 1, pp. 61–81.

Siggel, E. (1986) 'Technology transfers to developing countries through consulting engineers: A model and empirical observations from Canada', *The Developing Economies*, vol. 24, no. 3, pp. 229–250.

Stead, D. (2012) 'Best practices and policy transfer in spatial planning', *Planning Practice and Research*, vol. 27, no. 1, pp. 103–116.

Steelman, T.A. and Ascher, W. (1997) 'Public involvement methods in natural resource policy making: Advantages, disadvantages and trade-offs', *Policy Sciences*, vol. 30, no. 2, pp. 71–90.

Stenmark, D. (2002) Information vs. knowledge: The role of intranets in knowledge management, Proceedings of the 35th Hawaii International Conference on System Sciences, 7–10 January 2002, IEEE Press, Hawaii.

Stone, D. (1999) 'Learning lessons and transferring policy across time, space and disciplines', *Politics*, vol. 19, no. 1, pp. 51–59.

Stone, D. (2000) 'Non-governmental policy transfer: The strategies of independent policy institutes', *Governance*, vol. 13, no. 1, pp. 45–70.

Stone, D. (2004) 'Transfer agents and global networks in the 'transnationalization' of policy', *Journal of European Public Policy*, vol. 11, no. 3, pp. 545–566.

Svensson, R. (2007) 'Knowledge transfer to emerging markets via consulting projects', *The Journal of Technology Transfer*, vol. 32, no. 5, pp. 545–559.

Swainson, R. and de Loe, R.C. (2011) 'The importance of context in relation to policy transfer: A case study of environmental water allocation in Australia', *Environmental Policy and Governance*, vol. 21, no. 1, pp. 58–69.

Swyngedouw, E. (2005) 'Dispossessing H_2O: The contested terrain of water privatization', *Capitalism Nature Socialism*, vol. 16, no. 1, pp. 81–98.

Trott, P., Cordey-Hayes, M. and Seaton, R.A.F. (1995) 'Inward technology transfer as an interactive process', *Technovation*, vol. 15, no. 1, pp. 25–43.

UN (United Nations) (2002) Johannesburg Declaration on Sustainable Development, United Nations, New York.

UNCED (United Nations Conference on Environment and Development) (1992) Rio Declaration on Environment and Development, United Nations, New York.

UNECE (United Nations Economic Commission for Europe) (1998) Convention on Access to Information, Public Participation in Decision-making and Access to Justice in Environmental Matters, UNECE, Aarhus.

Vinke-de Kruijf, J., Augustijn, D.C.M. and Bressers, J.T.A. (2012) 'Evaluation of policy transfer interventions: Lessons from a Dutch–Romanian planning project', *Journal of Environmental Policy and Planning*, vol. 14, no. 2, pp. 139–160.

Vinke-de Kruijf, J., Hulscher, S.J.M.H. and Bressers, J.T.A. (2011) Knowledge transfer in international cooperation projects: Experiences from a

Dutch-Romanian project, Conference Paper, 5th International Conference on Flood Risk Management (ICFM5), 27–29 September 2011, Tokyo.

Von Korff, Y., Daniell, K.A., Moellenkamp, S., Bots, P. and Bijlsma, R.M. (2012) 'Implementing participatory water management: Recent advances in theory, practice, and evaluation', *Ecology and Society*, vol. 17, no. 1, art. 30.

Walker, J.L. (1969) 'The diffusion of innovations among the American states', *The American Political Science Review*, vol. 63, no. 3, pp. 880–899.

Ward, S.V. (1999) 'The international diffusion of planning: A review and a Canadian case study', *International Planning Studies*, vol. 4, no. 1, pp. 53–77.

Webler, T. (1995) '"Right" discourse in citizen participation: An evaluative yardstick', in O. Renn, T. Webler and P. Wiedemann (eds) *Fairness and Competence in Citizen Participation: Evaluating Models for Environmental Discourse*, Kluwer Academic Publishers, Dordrecht.

White, G.F. (1998) 'Reflections on the 50-year international search for integrated water management', *Water Policy*, vol. 1, no. 1, pp. 21–27.

Wolman, H. and Page, E. (2002) 'Policy transfer among local governments: An information–theory approach', *Governance*, vol. 15, no. 4, pp. 477–501.

World Bank (1994) *Governance, The World Bank's Experience*, The World Bank, Washington, DC.

WWAP (World Water Assessment Programme) (2006) *The United Nations World Water Development Report 2: Water, A Shared Responsibility*, UNESCO, Paris and Berghahn Books, New York.

WWAP (World Water Assessment Programme) (2012) *The United Nations World Water Development Report 4: Managing Water under Uncertainty and Risk*, UNESCO, Paris.

3 Contextual Interaction Theory for assessing water governance, policy and knowledge transfer

Hans Bressers and Cheryl de Boer

3.1 Introduction

The successful transfer and eventual implementation of a policy is heavily dependent on the relationship between the context from which it came and that to which it is transferred. In Chapter 2, Vinke-de Kruijf and Özerol elaborates on the numerous ways in which this relationship is understood and analysed by different streams of literature. This chapter will provide a thorough explanation of one theoretical lens, which has been used extensively to examine water governance and implementation related processes: Contextual Interaction Theory (CIT). The basic premise from which CIT begins is that context matters, and quite a lot. While Chapter 2 focuses on the processes associated with the transfer of a policy, CIT is used here to provide a way to understand the internal processes that occur once the governance actors have decided to move forward to the implementation of the chosen policy. The purpose of introducing such a theory here is that it provides an analytical framework that helps to assess whether a policy transfer contributes to or damages conditions for successful water management. This chapter goes beyond addressing the likelihood of policy transfers and focuses on the degree of success of such transfers. Once a policy transfer is in principle accepted, like water management based at the river basin scale or stakeholder participation, it will be confronted with an existing governance regime. This confrontation can provide resistance or simply encapsulate what is new rather than really change to adapt to the new concepts or principles. Further, if the existing arrangements do undergo change there is no guarantee that the ideas that worked so well in the originating context will be able to contribute to the quality of water governance in the new context in which they are being applied.

Within policy studies there are wide ranges of frameworks, theories and methodologies available that can be used by researchers and practitioners in order to better understand how, when and why the context surrounding policies and projects affects inputs, processes and outcomes. CIT places the characteristics of the actors involved in the policy process at the centre of the analysis and provides a framework within which the influences of the external context can be understood. Through a more detailed understanding of how, where and why these actor characteristics are influenced by the external context, one can begin to draw some lessons related to effective policy transfer in water governance.

In the next section we first explain some basics of the actor-centred orientation of this framework and how actor characteristics ultimately determine the success of any policy and thus policy transfer. Thereafter the layers of context that affect these actors are presented. Policy transfers can be viewed as interventions into these contexts that not only need to assume their place therein, but also create synergy with existing elements to produce tangible improvements.

3.2 Actor characteristics as the ultimate process setting

When policy processes are viewed themselves as social interaction processes that are ultimately driven by the actors involved, as they are in Contextual Interaction Theory, it makes sense to place them and their main characteristics central stage in any analytical model, and to build any further explanation of the course and results of the process from that simple starting point. Throughout the history of policy implementation research, hundreds of crucial success factors have been proposed and used to analyse various cases. This can be theoretically interesting when trying to carve out the impact of a single factor. Practitioners however deal with situations in which numerous factors are simultaneously present and where different combinations arise. Even in a rather simple model of fifteen contextual factors each having only two possible values there are some 30,000 different possible combinations of circumstances. That model is unworkable as an analytical tool (Goggin, 1986), and excessive for use in practical analysis.

CIT asserts that since interaction processes are human activities, all influences – including those created by policy instruments – will flow via the key characteristics of the actors involved. These three characteristics are the cognitions, motivations, and power and resources of the actors. This approach enables an initial distinction of the inner core factors that is far more parsimonious. Figure 3.1 outlines these factors (inputs and outputs of the process are not shown in this figure).

The characteristics of the actors in the process are the ultimate driving forces of the process, and not mere consequences of the 'arena' in which the process takes place (cf. Ostrom, 1999). The resources of the actors provide them with capacity to act and power in relation to other actors. Motivations and cognitions also play an important role in creating productive or non-productive settings for the process (see also further below). Resources become significant to the process when they are connected to cognitions and motivations. These three core actor characteristics represent different perspectives on social interaction processes and are exceptionally useful in explaining the dynamics of such processes (Owens, 2008).

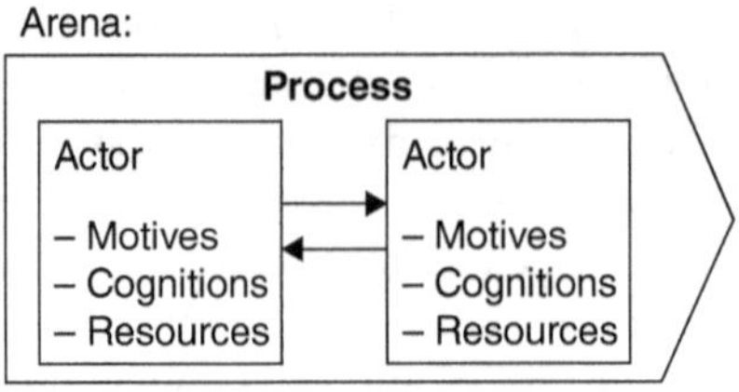

Figure 3.1 Process model with the actor characteristics used in Contextual Interaction Theory.

In this model peope constitute the 'actors', both individual and collective. Quite often however, these people represent organizations or groups and so often these organizations rather than the individual people are considered as the 'actors'. In Chapter 2, four types of actors are discerned that play a role in policy transfers, including supranational and international actors. In terms of the process, the relevant characteristics of representatives are often determined to such a large extent by the organization or group that they represent. As a result, a change of the individual actor is in most cases not likely to significantly change the setting of the process. 'Where one stands depends on where one sits' is an aphorism already cited by Allison (1971, p. 176) in his classic policy analysis. There is however also potential for the characteristics of individual people to have an impact in cases where they possess (or lack) diplomatic skills, creativity, motivation supported by personal enthusiasm, or where productive or destructive 'chemistry' exists between the individuals involved.

Summarizing this section: Contextual Interaction Theory explains the dynamics of social interaction processes and is intended to be simple and straightforward (Bressers, 2009). The theory's first main assertions are:

a Policy processes are not mechanisms, but social interaction processes between a set of actors (people, organizations). These policy processes include policy transfer processes and project realization.
b Many factors can influence the activities and interactions of these actors but only because and in as far as they change relevant characteristics of the involved actors.
c These characteristics are: their motives (which drive their actions), their cognitions (information held to be true, with which the situation is interpreted) and their resources (providing capacity and power) (see also Bressers, 2004).
d These three characteristics influence each other, but cannot be restricted to two or one without losing much insight (see Bressers, 2009).
e The characteristics of the actors shape the process, but are in turn also influenced by the course of and experiences in the process and can therefore gradually change during the process.

3.3 Layers of context and their relevance

The three main actor characteristics are not only intrinsic to the actors and influenced by the process, but are also influenced by many external factors from a multilayered context. Part of that context is the case-specific context. This involves factors like the characteristics of the geographical place where the project is realized (Kotzebue *et al.*, 2010). A special category is that of the case history consisting of previous decision-making and framing. This special type of case history affects the institutional arena for the process that influences which actors participate, to what extent and with what legal resources and expectations. A further layer of context is the structural context, with both the elements of governance and the relevant property and use rights (Bressers and Kuks, 2004). Next there is a less specified layer of wider contexts, within which the culture, and economic and technological developments and political system (Brynard, 2005, p. 659) are included.

The structural (governance) context related to for instance the national level is much more stable than the specific context, which applies to individual cases. The structural context will to a far lesser degree be influenced back by individual cases of policy transfer and implementation of the transferred options. In fact it is the essence of the difference between the specific and the structural context that the latter holds for, in principle, all similar cases and not only for any specific case. Nevertheless it too will gradually change in processes on a larger scale than the specific case, but

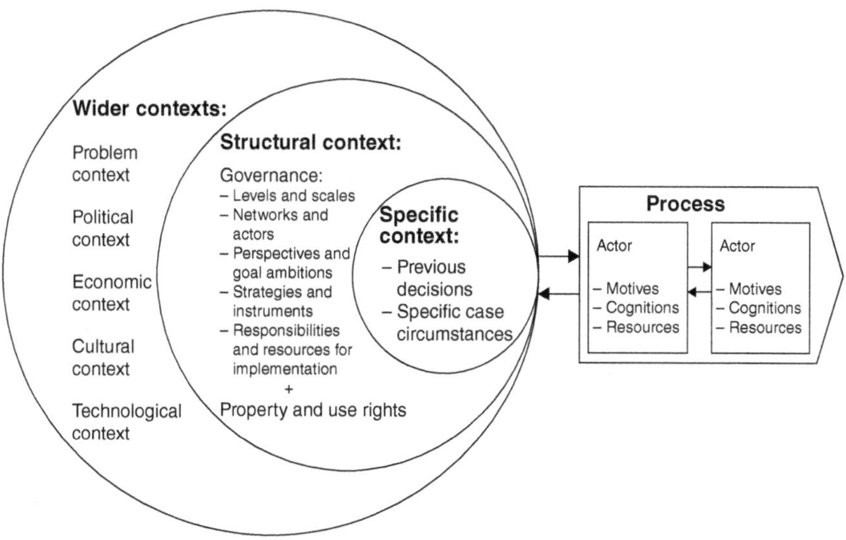

Figure 3.2 Layers of contextual factors for actor characteristics (source: Bressers, 2009).

with similar dimensions of motivational, cognitive and resource developments in response to external influences and internal frictions (Bressers and Kuks, 2003, p. 74–83; Costejà Florensa, 2003). These developments are not always coherent across the various elements of governance (compare Howlett and Cashore, 2007). The elements of governance also influence each other when new situations are encountered. For instance the degree of interconnectedness and cohesion of the network relations influence the characteristics of instruments in instrument selection processes (Bressers and O'Toole, 1998, 2005; Ligteringen, 1999).

The specific context is not entirely determined by the structural context. A lack of interconnectedness and cohesion of the network relations at the national governance level need not be replicated among the constellation of the actors involved. More generally in the context of a specific case, the structural relations between levels, actors, goals, instruments and resources will be adapted to the specific case in as far as the actors strive for this and it is possible. When the process continues for a longer period of time, or is a process in a series of similar processes that together give enough time, this adaptation is feasible. It can be worthwhile to attempt to build for instance better networked relations among the actors involved in this series of processes. Such collaborative policy networks at the case level can have important self-reinforcing characteristics (DeLeon and Varda, 2009). The strategies used by the actors involved for this 'metaprocess' are discussed later in this chapter.

The structural context also consists of the valid *property and use rights* system associated with land, water and other relevant resources (e.g. Ostrom, 1999). The relevant structural context – or 'regime' – is thus a combination of both public governance *and* property and use rights (Knoepfel and Nahrath, 2005; Knoepfel *et al.*, 2007) and the related exclusion of uses and access granted to users. Policy options for water management that can work well in their area of origin can be a complete mismatch in another context where the property and use rights related to the protection of private land are much stronger.

The 'five multiplicity aspects of governance' form an important part of the structural context (Bressers and Kuks, 2003). Governance is not used here as a normative concept or as a hypothesis of developments in government–society relationships (Howlett, 2011, pp. 7–10, see also Chapter 2 where the concept of 'good water governance' is elaborated upon), but as a neutral, yet enlarged understanding of the scope of the concept of 'policy'. The five elements of governance take the form of:

1 Scales and levels: governance assumes a multilevel character of all scales (jurisdictional, spatial, temporal etc.); in policy transfer processes often a transnational or international scale is included.
2 Actors and networks: governance assumes the multi-actor character of the relevant network(s).

3 Perceptions of the problem and goal ambitions (not just the objectives): governance assumes the multifaceted character of the problems and ambitions.
4 Strategies and instruments: governance assumes the multi-instrumental character of the strategies of the actors involved.
5 Resources and responsibilities for implementation: governance assumes the complex multiresource basis for implementation.

In the context of a policy being transferred from one water management regime to another, these various aspects can guide an analysis of how similar and different the important aspects of the two governance contexts are. This can provide important insights for understanding how the implementation process related to the transferred policy will be influenced by the governance regime. In addition to using these aspects to study the influences of the structural context, a number of qualities of the governance regime can also be analysed and be used to understand its influence on the interaction processes. These qualities, extent, coherence, flexibility and intensity, are described in the following sections.

Around the structural context there is yet another more encompassing *wider context circle* of political system, sociocultural, economic, technological development and problem contexts. Some cultural settings can for instance make hierarchical approaches less feasible, or make some degree of social control obsolete, (Schwarz and Thompson, 1990; Thompson *et al.*, 1990; Wildavsky, 1982).

3.4 Extent and coherence as requirements for effective water resource management

The structural context influences the process not only through its direct contents, but also though its *extent* and *coherence* (Knoepfel *et al.*, 2001, 2003; Bressers and Kuks, 2004). The extent is a quality of the regime that refers to its completeness. The coherence is a quality that expresses the degree to which the various elements of the regime are strengthening rather than weakening each other. The success of policy transfers is partly dependent on whether these qualities are improved or decreased by the transfer.

A regime increases its extent and consequently becomes more complex when more levels and scales are involved, more actors are involved, more perceptions of the problem and accompanying goals are involved, more instruments are part of the policy mix and more organizations share responsibilities for implementation. The most eminent feature of extent is however the gradual increase of the domain of the regime, which consists of the uses and users regulated by one or more parts of the regime. With this also comes an increase in relevant property and use rights. This is then viewed as an increase in the crucial variable of *extent.* Regimes with an insufficient extent are by definition weak as guardians of sustainable resource use, since some relevant parts of the domain are unregulated. Policy transfers tend to increase the extent of the governance regime, since a cognition that important aspects are lacking in relation to the water problems at stake can be a main driver to engage in such transfers.

Complexity, as a characteristic of the governance regime, need not be harmful to the overall water governance processes present in a particular case (de Boer and Bressers, 2011). Logically, growing complexity develops as an answer to real needs and issues. Societies in modern times have generally grown towards increased complexity. Increased populations, borders, overlaps, activities, rivalries etc. are a fact of our current living environments. A growing complexity in governance can be viewed as a natural adaptation to that development (Gerrits, 2008; Teisman *et al.*, 2009). Many external change agents, such as technological developments, add new scales, new actors, new problem perceptions, new instruments and new responsibilities to the existing ones.

While the term 'integration' is common in policy papers (e.g. those on 'integrated water resources management'), coherence is used here instead, for the reason that, in policy papers the term integration (e.g. in 'integrated water management') is used in a sense that implicitly or explicitly includes an increase in the domain of the regime (the extent increased to all relevant users and uses). Therefore, we believe that integration as it is used in the policy sphere is a combination of what we call *extent and coherence.* In Chapter 2, the inherent tension between those two aspects in the concept of integrated water resources management (IWRM) is described. For the sake of conceptual clarity and the possibility of adapting to the meaning of the term integration in policy practice, we use the terms extent and coherence when appropriate, and reserve 'integration' for the combination of the two.

When more than one layer of government influences the management of the same water or other natural resource (as is often the case), then coherence means, among other things, that the activities of these layers of government are recognized as mutually dependent and influencing each other's effects. Likewise if more than one governance scale is relevant the interaction effects between those scales should be considered. When more than one actor (stakeholder) is involved in the policy, coherence means that there is a substantial degree of interaction in the policy network; preferably productive interaction providing coordination capacity. When more than one use or user is causing the problem of unsustainable resource use for example, coherence means that the various resulting objectives are analysed in one framework so that deliberate choices can be made if and when goals and/or uses are conflicting. When the actors involved have problem perceptions that start from different angles, coherence means that they are capable of integrating these to such an extent that a common ground for productive deliberation on ambitions is created. The same holds for instrumental strategies that are used to attain the different objectives, as well as for the different instruments in a mix to attain one of these objectives. Coherence of the organization of the implementation process means that responsibilities and resources are coordinated, or the responsible actors themselves are coordinated.

With more coherence in the public governance component of the regime, the goals of the actors involved in the process are less likely to be in discord. All elements of a more coherent regime can be assumed to contribute to a lowered degree of experienced uncertainty, an increase in information exchange, and a lower degree of distrust. This implies that a more coherent public governance component of the regime can outperform a regime with an equal degree of extent, yet more fragmentation. While policy transfers are assumed to generally increase the extent and thereby the complexity of the governance regime, the coherence with the existing elements of governance could actually be lowered rather than raised. In such cases, the overall effectiveness of water management will not improve, and could be in danger of deteriorating. Hence the degree of ultimate success of the policy transfer is also dependent on the degree to which it can be introduced in such a way that it improves rather than decreases governance coherence.

This leads to some further assertions:

f Regimes with a deficient extent are more likely to lead to a degradation of water resources or an inability to protect the ecological functions of the resource, than are regimes with a larger extent.

g Regimes with a large extent, but with low coherence are more likely to lead to degradation of natural resources or an inability to protect the ecological functions of the water resource, than regimes with a similar extent but a higher degree of coherence.

These last assertions have been tested as hypotheses and generally confirmed in a six country, twenty-four cases study on tributary river basin management (Bressers and Kuks, 2004). However the relation between the growth of the extent of the regime and the improvement of sustainability estimates proved to be rather weak and hardly significant. The relation between the general assessment of regime change towards more integration (extent and coherence taken together) and the assessment of sustainability improvements is however much stronger (Spearman's $rho = 0.533$, $p = 0.004$). Of the separate regime aspects, by far the most important factor was the coherence of public governance. It correlated even more strongly with the assessment of sustainable resource use than the general regime change ($rho = 0.635$). Another example confirming the CIT relationships between extent and coherence of governance regimes, the motivation, cognitions and resources of actors and the policy results, can be found in a separate study on Greece and the implementation of the EU Water Framework Directive (Kampa, 2007).

3.5 Governance flexibility and intensity as requirements for adaptive management

Extent and coherence have been shown to be important regime qualities in more or less steady state situations or when one wants to compare the before and after examples of policy transfers in water management. Following the general acceptance of the integrative ambitions of IWRM it was observed that the increased complexity leads to uncertainty and unpredictability and that water management thus needs to be prepared to avoid threats and use opportunities as they arise. This includes trying not only to see the reality as a field of obstacles, but also as a terrain of potential – often unexpected – opportunities and being adaptive enough to use every 'window of opportunity' to bring the ultimate purpose closer to realization. Sometimes the above is called Adaptive Water Management (AWM) (Lulofs and Bressers, 2010), though others emphasize the adaptive capacity of the water system itself as a core issue of this approach (Pahl-Wostl *et al.*, 2005). In a highly dynamic process situation in which success depends on quick and timely adaptive action, new regime qualities become important for supporting water management processes. Thus also the success of transferred policy options is partly dependent on the degree to which such governance regime qualities are reinforced or hampered. This section provides a description of the value added when flexibility and intensity are added as regime qualities when studying complex and dynamic policy processes and the role of policy transfers therein.

First, flexibility is important as it indicates to what degree the relevant actors have formal and informal liberties and stimuli to act. The attention given to the flexibility of the regime is approached on the basis of the documentation about the regime elements, as well as from a bottom up perspective. Success and failure factors of the actual projects can be rightfully attributed to inflexibilities in provincial, national and ultimately supranational policies. What is also interesting is how the latter are moderated by national and subnational regimes.

Flexibility as defined in de Boer and Bressers (2011, p. 92) is 'the degree to which the regime elements support and facilitate adaptive actions and strategies in as far as the integrated (among others multi-sectoral) ambitions are served by this adaptiveness'. Consequently it is also the degree to which hindrances for such adaptive behaviour are avoided. The addition 'in as far as …' is needed to discern governance and policy that is just weak from a genuine attempt to make the most of the situation. The term 'integrated (among others multi-sectoral) ambitions' refers here to the integrated multifunctionality of the various interests that exist in the project or policy arena. In cases of water policy transfer, there are likely to be many other non-water-related interests that will be affected throughout the policy process and as a result of the resource use arrangement that is in place following implementation. Thus, the context will determine to what extent new and existing uses need to be reconciled during this process.

The value of flexibility must be understood in connection with the level of the other qualities also present in the regime. Flexibility needs at least a certain degree of coherence to be built upon. Otherwise, when it is just extensive discretion and self-reliance for water managers there is a high risk that a fragmented and a weak form of water management would result.

The flexibility of the regime can be described in terms of the five elements of governance. The following paragraphs elaborate on flexibility in terms of the five aspects of governance (see also de Boer (2012) for further development of this concept):

1 A regime is more flexible in as far as the relationships between the levels and scales involved are more based on decentralization of power, without upper levels withdrawing support. This is closely related to empowering rather than controlling relations, and thus on trust.
2 A similar feature describes flexible regimes in terms of actor relations in the policy network. The combination of giving leeway to each actor group to optimize its contribution to the whole programme while still viewing the programme as a joint effort qualifies as flexibility.
3 In terms of general problem perception and goal ambitions flexibility implies that these in their variety are not only integrated into a sort of common denominator (like with coherence), but also that these mixtures are allowed to be different in emphasis according to the opportunities of the context in the various concrete situations. This implies some acceptance of uncertainty and openness to emergent options, which again relates to trust.
4 The instruments and their combinations in policy strategies or mixes are more flexible in as far as means from different sources (like public policies and private property rights) may be used as well as indirect means (here relating to opening or improving options for the use of means that more directly serve the goals) are available and allowed to be used.
5 The flexibility of the organization responsible for the implementation – the responsibilities and resources given by the policy programme(s) – can be measured by the discretion available to pool resources like funds and people with those of others to serve integrated projects and to be held accountable on the basis of the balanced virtues of the achievements (as in an integrated project), rather than on the basis of separate performance criteria.

There is often a highly dynamic and change-oriented nature to many policies that aim to improve general water governance processes or the use of water resources. This is the case since water management often influences many societal issues related to health, economic development, environmental resilience etc. As such, there is yet another regime quality that can be influential for the practical process of implementing water related policies. That is the obvious, but no less important aspect of intensity. *Intensity* is 'the degree to which the regime elements urge changes in the status quo or in current developments' (de Boer and Bressers, 2011, p. 93). The 'amount of change' is thereby measured in analogy with Newton's 'law of inertia', respect to the degree of energy it takes to produce the change. In systems theory, induced changes will typically meet negative feedback loops, weakening their impact, while in some cases positive feedback loops creating dynamics for permanent change are also conceivable (True *et al.*, 1999; Bressers and Lulofs, 2009). In policy studies terms, intensity is related to the size of the task that is required to create new dynamics by creative cooperation, or conflict. Conservative motivations often need to be changed or overruled by power. Cognitive boundary judgements regarding the issues at stake require broadening. New resources need to be made available and combined with old ones. In other words: with more intensity the urge to use clever adaptive strategies to deal with and change the setting of the process increases.

The five elements of governance can also be used to highlight the important intensity related aspects of the governance regime. By doing so we find that:

1 Intensity is greater in as far as upper levels are more deeply involved.
2 Actors that are also powerful in other domains are more deeply involved in the relevant policy network for the issue at stake.
3 The issue plays a larger role in the public debate leading to a greater openness to try to push developments away from a business-as-usual track (thus with more ambitious goals).
4 The instruments made available to be used include more interventionist ones.
5 The amount of resources made available for implementation is larger. (See also de Boer (2012) for further development of this concept.)

We acknowledge that there is an implicit potential tension between this 'quality' of the regime and the previous one of flexibility. This is related to the eternal dilemma of the 'quest for control' involving distrust, versus the 'learning while doing approach' involving trust. When an actor wants to achieve a great deal of change this can increase the distrustful tendency to try to control the related or influencing processes from the top down, which leads to a decrease in the flexibility. However in complex and dynamic situations decreasing flexibility is often detrimental since arising opportunities are missed or not created and developing obstacles often cannot be foreseen. However when there is sufficient trust in the implementers' motivation to genuinely work on the matter, flexibility in the governance regime can be an important resource.

When the change striven for is multidimensional, e.g. involving multiple policy sectors, synergies should be welcomed and trade-offs accepted when necessary (de Boer *et al.*, 2011). While in its definition 'flexibility' is shielded from mere discretion that would also accept implementers doing nothing, there remains a natural limit to flexibility as a positive force. When multiple policy sectors are involved in the water management process, an integrated vision is helpful for guiding the process, preventing extreme imbalances (coherence) or the exclusion of essential sectors (extent). Inevitably there will be some limitations to flexibility induced by this. Here the regime quality of flexibility is restricted to a certain degree by the regime quality of (multi-sectoral) interregime extent and coherence.

Some last assertions of Contextual Interaction Theory specified here are:

h While the *extent* and *coherence* are crucial qualities of the structural context when stabilization and protection of a resource is attempted through the policy intervention, other qualities should be added when change and the creation of new resources are desired.

i The first additional quality is the *flexibility* of the regime; the degree to which the regime allows and facilitates the case-specific variation and boundary spanning strategies of actors needed for adaptive management in as far as the change ambitions are served by this adaptiveness. Under the conditions of sufficient motivation of the implementers and sufficient interregime extent and coherence more flexibility will lead to better adaptive strategies and thereby to improved results.

j The second additional quality is the *intensity*: to what degree is the change striven for a deviation from 'business as usual'? The greater the intensity, the more resistance that will have to be overcome (negative feedback loops), but sometimes also more enthusiasm can be provoked (positive feedback loops).

3.6 Adaptive strategies to deal with complex and dynamic settings

The use of adaptive management is presented as an essential remedy for the problems seen around the world related to access to and management of water (Pahl-Wostl *et al.*, 2005; Bressers and Lulofs, 2010b). This adaptive management is itself however not a process without its own internal struggles and various strategies exist with which to accomplish its end goals. The strategies that develop can be a response to unsatisfactory processes, though with experienced actors they can also be preventative, attempting to avoid an unproductive setting before it becomes a fixed context for the ensuing process. Such strategies rarely have a 'the more the better' character. In fact their (degree of) application needs to be continuously balanced with the evolving threats and opportunities of the context, that often itself is in flux for more reasons than deliberate strategies of the actors involved. These strategies can involve more than just the motivations, cognitions and resources concerning the issues at stake. The actors can also try to change the specific context of the process. Sometimes it is possible to bring in new actors or exclude existing ones or try to redefine the process and its relevant issues differently to shift it to another arena with a (partially) different governance context (see for an elaboration Bressers and Lulofs, 2010a, pp. 27–30). Some examples which have been observed are:

1 adding new actors; these can include policy brokers or 'inviting oneself' to consultations among other actors on new plans, or more generally keeping regular interaction with other potentially relevant actors even when no immediate issue is calling for attention, this way providing better network relationships not on the level of the general governance context but much closer to the level of the specific case;
2 creating new arenas by adding new meeting points such as through the choice of a particular administrative setting (e.g. a voluntary process or under the institutional setting of a certain law) for the process when possible or by installing working groups or committees;
3 creating new cognitions by introducing new information, spreading information and perceptions by new catchy key words and metaphors, involving the media etc. (compare van Buuren, 2006);
4 creating new motivations by creating salience among others through luring with resources, or by promoting with positive intermediate results; and
5 adding new resources and power bases for instance by exchanging relatively flexible ones (like money) in advance for relatively fixed or scarce ones, such as land ownership.

The use of strategies is often not deliberate or precise. Understanding of strategic options must be accompanied by the presence of the necessary and efficient tools. These options also require skilful actors to enable their meaningful exploitation (Hermans, 2005). In complex systems, many developments are 'emergent'; quite unpredictable consequences of a multiplicity of factors and circumstances. Consequently, the use of strategies is often spontaneous and time-pressured. Moreover, it can be quite wise to simultaneously use diverse strategies and include a degree of 'redundancy' in order to create fall-back options when one of the strategies does not have its intended or desired effect, as the literature also states about water management institutions (Constanza *et al.*, 2001; Low *et al.*, 2003; Ostrom, 2005). While it is important to keep various water management options open as long as possible, in order to provide a sufficient degree of accountability towards elected representatives, it is necessary to work within an agreed dynamic framework for various stages of action (Koppenjan *et al.*, 2009). Such accountability need not be at the expense of organizational capacity when the performance criteria are under reflexive adjustment and do not remain fixed regardless of new situations (Pires, 2011). 'Good water management will only become a reality once we recognize that water managers have complex duties' (Figuères *et al.*, 2003).

Vinke-de Kruijf and Özerol discuss in Chapter 2 some insights from the realm of technology transfer that were later applied to the realm of water resources management. When we look at the strategies that actors in dynamic situations apply to themselves internally, by trying to prepare themselves to take advantage of these opportunities for improved interaction in complex settings, we see these concepts as connecting these two aspects of the policy transfer and water management implementation process. Internal strategies are seen as an attempt to increase the actor's receptivity. Jeffrey and Seaton (2004) coined the term receptivity as not only dependent on the degree of exposure to new knowledge, but also more specifically on the way the actor can associate and exploit new knowledge around existing knowledge, activities and objectives. This requires that the actor 'lets the outside come in', and can open and regroup their understandings to include reckoning with the new knowledge. Thus, receptivity is not a form of passiveness or weakness. It is a form of alertness and openness towards the context that enables well-targeted innovative and adaptive action by self-confident people and organizations.

Receptivity tends to play a major role in recognizing the opportunities that can be seen due to the enlarged domain perception and then used to create synergies with the activities of other actors. If both parties perceive potential synergy, meaning that they see joint chances for cooperation, boundary spanning is more likely to create productive linkages (Bressers and Kuks, 2004, pp. 259–262). This perception in turn reinforces the degree of openness towards enlarged domain boundaries. If one or both parties consider the situation purely as rivalry or even mutually exclusive, one might even observe attempts to reinforce existing boundaries or bring up new boundaries in order to keep domains apart or separate them. Jeffrey and Seaton (2004) discern four phases of receptivity that can also be seen as permanent variables: awareness, association, acquisition and application (see also Bressers, 2011). In the cognitive system, these can be linked to, respectively, the observations, the filtering through frames of reference (including boundary judgements on what belongs to the subject of the process at stake and what not), the interpretations of reality and the impacts of the cognitive system on motivation, capacity and the process itself.

For the purposes of understanding the specific aspects of implementing a transferred water policy in a multidisciplinary and functional context, we use an expanded definition of receptivity:

> the ability to combine new information with existing cognitions, to recognize new goals as matching existing motivations or the values behind them and to recognize the opportunities of new resources or combinations with existing resources to optimize their capacity and power
>
> (De Boer and Bressers, 2011, p. 90)

A greater likelihood for productive interaction processes under one of the various combinations of motivations, cognitions and resources of the actors involved might result from an increase in such receptivity.

Receptivity can be both a quality of people and organizations. While we typically describe receptivity as an organizational characteristic, we acknowledge the importance of skilled, experienced and open-minded project managers that fulfil important roles as 'boundary spanners'. The receptivity of an organization is partly important to enable them to fulfil this function.

Like the arena and constellation of actors and their characteristics, the receptivity of an organization is not entirely fixed and can thus be altered in the course of time by external factors and deliberate *internal strategies.* It is crucial that an organizational philosophy is oriented towards external cooperation for instance through rewarding project managers that are communicative, flexible and entrepreneurial.

Summarizing this section, some further assertions in CIT are:

k The rules of the game that e.g. provide or restrict resources are often not static but themselves subject to change partly by *external strategies* by actors in the process unless they are firmly fixed by the regime. The same holds for the actor constellation in the process.

l The setting of actor characteristics that impacts on the course and result of the process is not only dynamic due to external factors, but can also to some extent be manipulated by clever *external strategies* of the actors during the process (these are often forms of 'boundary spanning' – see Bressers and Lulofs, 2010b).

m Since adaptive boundary spanning strategies often require concerted actions by more than one individual person, this also draws attention to the internal organization of the actor ('actors' in most analyses are in fact 'corporate actors': organizations or parts of organizations). Here the *receptivity* of these actors and actor organizations is relevant.

n This receptivity can also be positively influenced by *internal strategies* of actor organizations, which promote continuous learning, consciously dealing with uncertainty, and stimulating mutually supportive intraorganizational relations.

3.7 Concluding remarks: contextual factors set the conditions for policy and knowledge transfers

Understanding the various impacts from various perspectives and influencing factors – particularly in the complex policy realm of water governance – is highly dependent on context. The task stressed here is to develop an approach which assists us in knowing 'what works, where, when and how in transferring lessons in water management?' By being open to a large variety of contextual factors, but nonetheless channelling them through a limited number of 'core characteristics' that build a deductive explanatory framework, the approach that is provided in this chapter strives to achieve an understanding of the water management implementation processes which is simultaneously concise and more generally applicable. This framework supports an assessment of the degree to which policy transfers can be viewed as improvements that contribute to, rather than hamper, successful water management.

Reasons for using such a framework include being able to see the influence of different characteristics which arise from varying contexts and how they play a prominent role in the final product without attempting to cancel out the important local effects. The context is not only important in as far as how it impacts the original intent of the policy, but also gives it a place of importance in the expectation of the more widely defined preferred outcome. This form of utility is key to understanding the process and influences on the given actors involved in the transfer and implementation processes. Importantly, this model can also be used to describe the various interactions and influences between the actor (organization) of interest and other actors, regime contents and qualities under dynamic conditions and can focus on the impacts of these dynamics.

The policy process as seen to occur between the external context and the target actors is explored to show how the receptivity (characterized by awareness, association, acquisition, adoption) of the target organization can work as a catalyst to align the external and internal contexts to enable successful implementation of transferred policy options. Changes in context occur continually in any policy process due to the complexity of the regime or the dynamic nature with which actors interact.

The model developed in this chapter is used by a number of authors in the following chapters of this book in their explorations of particular experiences in knowledge and policy transfer in the field of water governance. It is thus provided here as a basis upon which different cases can be compared, however a number of the chapters make use of other appropriate models and frameworks. We have encouraged the use of these different approaches in this book and provide some conclusions in the final chapter about the value of having a number of different lenses from which to view contemporary water governance practices as well as the plethora of real-world case examples. The following chapters in this book provide insights on cases where policy and knowledge transfers are being undertaken as a way of addressing global and local water resource use and management challenges. This chapter presents a way of thinking about what happens when these policies take hold of resources and attention through more local implementation actions and water management projects.

References

Allison, G.T. (1971) *Essence of Decision: Explaining the Cuban Missile Crisis*, Little, Brown and Company, Boston, MA.

Bressers, Hans (2004) 'Implementing sustainable development: How to know what works, where, when and how', in William M. Lafferty (ed.) *Governance for Sustainable Development: The Challenge of Adapting Form to Function*, Edward Elgar, Cheltenham, Northampton MA, pp. 284–318.

Bressers, Hans (2009) 'From public administration to policy networks: Contextual interaction analysis', in Stéphane Narath and Frédéric Varone (eds) *Rediscovering Public Law and Public Administration in Comparative Policy Analysis: A Tribute to Peter Knoepfel*, Presses polytechniques, Lausanne, pp. 123–142.

Bressers, H.T.A. and Kuks, S.M.M. (2003) 'What does "governance" mean? From conception to elaboration', in H. Bressers and W.A. Rosenbaum (eds) *Achieving Sustainable Development: The Challenge of Governance across Social Scales*, Praeger Publishers, Westport, CT.

Bressers, Hans and Kuks, Stefan (eds) (2004) *Integrated Governance and Water Basin Management: Conditions for Regime Change and Sustainability*, Kluwer Academic Publishers, Dordrecht/Boston, MA/London.

Bressers, Hans and Lulofs, Kris (2009) 'Explaining the impact of the 1991 and the 2000 firework blasts in the Netherlands by the core of five policy change models', in Giliberto Capano and Michael Howlett (eds) *European and North American Policy Change: Drivers and Dynamics*, Routledge, Abingdon and ECPR, New York, pp. 15–42.

Bressers, Hans and Lulofs, Kris (2010a) 'Analysis of boundary judgments in complex interaction processes', in Hans Bressers and Kris Lulofs (eds) *Governance and Complexity in Water Management*, Edward Elgar, Cheltenham, pp. 17–32.

Bressers, H. and Lulofs, K. (eds) (2010b) *Governance and Complexity in Water Management*, Edward Elgar, Cheltenham.

Bressers, Hans and O'Toole, Laurence J., Jr. (1998) 'The selection of policy instruments: A network-based perspective', *Journal of Public Policy*, vol. 18, no. 3, pp. 213–239.

Bressers, Hans and O'Toole, Laurence J., Jr. (2005) 'Instrument selection and implementation in a networked context', in Pearl Eliades, Margaret M. Hill and Michael Howlett (eds) *Designing Government: From Instruments to Governance*, McGill-Queen's University, Montreal/Kingston/London/Ithaca, NY, pp. 132–153.

Bressers, N. (2011) 'Co-creating innovation: A systemic learning evaluation of knowledge and innovation programmes', PhD thesis, Erasmus University Rotterdam.

Brynard, P. (2005) 'Policy implementation: Lessons for service delivery', *Journal of Public Administration*, vol. 40, no. 4.1, pp. 649–664.

van Buuren, Arwin (2006) *Competente besluitvorming (Competent Decision-Making)*, Erasmus University, Rotterdam.

Constanza, R., Low, B., Ostrom, E. and Wilson, J. (2001) *Institutions, Ecosystems and Sustainability*, Lewis, New York.

Costejà Florensa, Meritxell (2003) Institutional dynamics: First steps toward developing a logic sequence for studying institutional change, Paper conference of Workshop in Political Theory and Policy Analysis, Bloomington, IN, 13–15 December.

de Boer, Cheryl (2012) 'Contextual water management: A study of governance and implementation processes in local stream restoration projects', PhD thesis, University of Twente, Enschede, pp. 1–265.

de Boer, C. and Bressers, H. (2011) *Complex and Dynamic Implementation Processes: the renaturalization of the Dutch Regge River*, University of Twente with the Dutch Water Governance Centre, Enschede.

de Boer, Cheryl, Bressers, Hans and Kuks, Stefan (2011) 'Coordination of policies and governance: Regime requirements in Dutch freshwater management', *Policy Quarterly*, vol. 7, no. 4, pp. 3–9.

DeLeon, Peter and Varda, Danielle (2009) 'Toward a theory of collaborative policy networks: Identifying structural tendencies', *Policy Studies Journal*, vol. 37, no. 1, pp. 59–74.

Figuères, Caroline, Rockström, Johan and Tortajada, Cecilia (2003) 'Conclusion: The way forward', in Caroline Figuères, Johan Rockström and Cecilia Tortajada (eds) *Rethinking Water Management: Innovative Approaches to Contemporary Issues*, Earthscan, London, pp. 228–236.

Gerrits, Lasse (2008) *The Gentle Art of Coevolution*, EUR, Rotterdam.

Goggin, Malcolm L. (1986) 'The 'too few cases/too many variables' problem in implementation research', *Political Research Quarterly*, vol. 39, no. 2, pp. 328–347.

Hermans, Leon M. (2005) *Actor Analysis for Water Resources Management*, Eburon, Delft.

Howlett, Michael (2011) *Designing Public Policies: Principles and Instruments*, Routledge, Abingdon.

Howlett, Michael and Cashore, Ben (2007) The dependent variable problem in assessing policy change: Re-conceptualizing the orthodox understanding of policy dynamics, paper ECPR Joint Sessions, 7–12 May, Helsinki.

Jeffrey, P. and Seaton, R.A.F. (2004) 'A conceptual model of "receptivity" applied to the design and deployment of water policy mechanisms', *Environmental Sciences*, vol. 1, no. 3, pp. 277–300.

Kampa, Eleftheria (2007) *Integrated Institutional Water Regimes: Realisation in Greece*, Logos, Berlin.

Knoepfel, Peter and Nahrath, Stéphane (2005) *Sustainable Management of Natural Resources: From Traditional Environmental Protection Policies Towards Institutional Natural Resource Regimes (INRR)*, Idheap, Chavannes-Lausanne.

Knoepfel, Peter, Kissling-Näf, Ingrid and Varone, Frédéric (2001) *Institutionelle regime für natürliche ressourcen*, Helbing & Lichtenhahn, Basel.

Knoepfel, Peter, Kissling-Näf, Ingrid and Varone, Frédéric (eds) (2003) *Institutionelle Ressourcenregime in Aktion/Régimes institutionnels de ressources naturelles en action*, (*Institional Resource Regimes in Action*), Helbing & Lichtenhahn, Basel/Genf/München.

Knoepfel, Peter, Nahrath, Stéphane and Varone, Frédéric (2007) 'Institutional regimes for natural resources: An innovative theoretical framework for sustainability', in Peter Knoepfel, *Environmental Policy Analyses: Learning from the Past for the Future: 25 Years of Research*, Springer, Berlin, pp. 455–506.

Koppenjan, Joop, Kars, Mirjam and Van der Voort, Haiko (2009) 'Vertical politics in horizontal policy networks: Framework setting as coupling arrangement', *Policy Studies Journal*, vol. 37, no. 4, pp. 769–792.

Kotzebue, J.R., Bressers, H.T.A. and Yousif, C. (2010) 'Spatial misfits in a multi-level renewable energy policy implementation process on the Small Island State of Malta', *Energy Policy*, vol. 38, no. 10, pp. 5967–5976.

Ligteringen, Josee (1999) 'The feasibility of Dutch environmental policy instruments', PhD thesis, University of Twente, Enschede.

Low, B., Ostrom, E., Simon, C. and Wilson, J. (2003) 'Redundancy and diversity: Do they influence optimal management?' in F. Berkes, J. Colding and C. Folke (eds) *Navigating Social-Ecological Systems*, Cambridge University Press, Cambridge, pp. 83–105.

Lulofs, Kris and Bressers, Hans (2010) 'Innovations in water management requiring boundary spanning: Roots and concepts', in H. Bressers and K. Lulofs (eds) *Governance and Complexity in Water Management: Creating Cooperation Through Boundary Spanning Strategies*, Edward Elgar Publishing, Cheltenham/ Northampton, MA, pp.1–16.

Ostrom, Elinor (1999) 'Institutional rational choice: An assessment of the institutional analysis and development framework', in Paul Sabatier (ed.) *Theories of the Policy Process*, Westview Press, Boulder, CO, pp. 35–71.

Ostrom, Elinor (2005) *Understanding Institutional Diversity*, Princeton University Press, Princeton NJ.

Owens, Katharine (2008) 'Understanding how actors influence policy implementation: A comparative study of wetland restorations in New Jersey, Oregon, The Netherlands and Finland', PhD thesis, University of Twente, Enschede.

Pahl-Wostl, C., Downing, T., Kabat, P., Magnuszewski, P., Meigh, J., Schlueter, M., Sendzimir, J. and Werners, S. (2005) *Transition to Adaptive Water Management*, NeWater Working Paper X, Institute of Environmental Systems Research, University of Osnabrück, Osnabrück.

Pires, Roberto R.C. (2011) 'Beyond the fear of discretion: Flexibility, performance, and accountability in the management of regulatory bureaucracies', *Regulation and Governance*, vol. 5, no. 1, pp. 43–69.

Schwarz, Michiel and Thompson, Michael (1990) *Divided We Stand: Redefining Politics, Technology and Social Choice*, Harvester Wheatsheaf, New York.

Teisman, Geert, Gerrits, Lasse and Buuren, Arwin van (2009) 'An introduction to understanding and managing complex process systems', in Geert Teisman, Lasse Gerrits and Arwin van Buuren (eds) *Managing Complex Governance Systems*, Routledge, New York, pp. 1–16.

Thompson, Michael, Ellis, Richard and Wildavsky, Aaron (1990) *Cultural Theory*, Westview Press, Boulder, CO.

True, J.L., Jones, B.D. and Baumgartner, F.R. (1999) 'Punctuated equilibrium theory: Explaining stability and change in American policy making', in Paul A. Sabatier (ed.) *Theories of the Policy Process*, Westview Press, Boulder, CO, pp. 97–115.

Wildavsky, Aaron (1982) 'The three cultures: Explaining anomalies in the American welfare state', *Public Interest*, no. 69, pp. 45–58.

Part I

Transferring established knowledge

4 How contextual factors influence the effectiveness of international projects

The case of Dutch-funded flood risk management projects in Romania

Joanne Vinke-de Kruijf, Carmen Teodosiu, Hans Bressers and Denie C.M. Augustijn

4.1 Introduction

River floods are one of the most threatening natural risks and have many negative consequences for humans and human society (Samuels *et al.*, 2006). Also in Europe, floods have been increasing and are currently the most common natural disaster (EEA, 2008). International organizations expect that the frequency and intensity of floods will only grow further due to climate change and the increase of population and economic activities in flood-prone areas (WMO, 2009). Within this context, international efforts aiming at flood risk reduction are growing. An example of such effort is the Associated Programme on Flood Management that promotes the concept of Integrated Flood Management (based on the concept of Integrated Water Resources Management (IWRM), see Chapter 2) and supports the implementation of actions that follow this principle (APFM, 2012). In a European context, best practices on flood management were recently translated into a framework directive for the assessment and management of floods (European Directive 2007/60/EC). While floods are common problems, floods have diverse causes and countries also have different capacities to deal with them (Van Alphen and Lodder, 2006). As Chapter 2 explains, universal remedies or blueprints for effective water management do not exist. This means that any international transfer of knowledge should be done with care and only after careful consideration of the context-specific conditions and circumstances in which knowledge was developed. While many scholars acknowledge the importance of context in water management, little is known about how contextual factors inhibit or promote knowledge transfer. This chapter aims to address this research gap by asking: how do contextual factors influence the effectiveness of water projects that involve international knowledge transfer?

In this chapter, knowledge transfer projects are defined as interactive processes in which actors of two or more different countries share and acquire knowledge for the purpose of using it for case-specific objectives (Vinke-de Kruijf *et al.*, 2011). This definition emphasizes that an effective knowledge transfer includes the application of knowledge. In this sense, knowledge transfers are very similar to policy implementation processes. We therefore assume that the course and outcomes of knowledge transfer projects can be understood using the Contextual Interaction Theory (see Chapter 3). Empirically, this chapter builds upon the experiences of three projects that involved the transfer of knowledge about flood risk management from the Netherlands to Romania and were implemented with the financial support of the Dutch government. This implies that the projects under study are examples of transfers that are driven by the proactive attitude of a transferring country to export its knowledge (Stone, 1999). We therefore choose to assess the effectiveness of these projects from a Dutch policy perspective and define effectiveness as the extent to which a project contributes to the solving of flood-related problems in Romania and generates follow-up projects for the Dutch water sector (Vinke-de Kruijf *et al.*, 2012).

The structure of the chapter is as follows. First, we introduce our main theoretical concepts and case study methodology. We then introduce the projects (including the context) of the case studies. Subsequently, we describe the motivations, cognitions and resources in relation to these projects. Following this analysis of actor characteristics, we analyse and discuss the influence of contextual factors on the cases. The last section presents our conclusion about how contextual factors influence the effectiveness of international knowledge transfer projects.

4.2 Theoretical concepts and methodology

Contextual Interaction Theory as conceptual framework

As explained in the introduction, knowledge transfer projects are in this chapter conceptualized as policy implementation processes. Inspired by the Contextual Interaction Theory, we assert that such processes are shaped by the characteristics, i.e. the motivations, cognitions and resources, of the actors involved. Therefore, the following questions are central in our analysis:

- What are the substantive or procedural reasons (motivations) of actors for being involved and taking certain actions?
- What are the perceptions (cognitions) of actors about the relevance of the project, the urgency, nature and meaning of the problem at stake and potential solutions?
- What is the capacity of actors to act (financial resources, human resources and knowledge) and power to get things done (institutional resources)? (Bressers and Kuks, 2004; Owens, 2008; Vinke-de Kruijf, 2011b).

In an interactive process, actor characteristics are in dynamic interaction with each other and tend to change. These changes can form a basis to attain ultimate outcomes (problem-solving and follow-up projects in this case). According to Vinke-de Kruijf *et al.* (2012), the likelihood of realizing ultimate outcomes especially depends on the development of a joint motivating goal, the creation of a negotiated knowledge base (i.e. knowledge that is agreed upon and valid) and the mobilization of necessary resources (by pooling them if necessary). In addition, positive relational experiences increase mutual trust and therefore improve the chances on future cooperation, but are not analysed in this chapter.

While the course and outcomes of an actor-interaction process result from the dynamic interaction between actor characteristics, they are also embedded in wider, structural and project-specific contexts (see Chapter 3). Although we acknowledge that contextual factors only influence a knowledge transfer process insofar as they change the characteristics of the actors involved, we consider an analysis of contextual conditions, circumstances and differences to be beneficial, since such an analysis helps to understand why actors possess or develop certain characteristics. Literature shows, for example, that transfers tend to be easier between countries that are similar in terms of resources (Rose, 1993). A comparative analysis highlights that participatory approaches in water management are more effective in some countries than in others due to institutional factors (Mostert *et al.*, 2007). Another study shows that actors tend to implement proposed reforms that are socially acceptable or economically attractive while neglecting reforms that are incompatible (Sehring, 2009). These examples show that transfers are influenced by cultural, institutional and historical configurations (Rose, 1993). While research shows that context influences policy transfers, contextual factors do not necessarily determine the success of a transfer. Comparative research shows, among others, that the effectiveness of a transfer is not correlated with contextual similarity but rather depends on whether transfers are supported by actors who are 'pulling in', i.e. whether powerful actors are convinced that the transfer is useful and have a strong desire to change things (Kroesen *et al.*, 2007). This chapter discusses two ways in which contextual factors are likely to influence the effectiveness of a knowledge transfer process. First, actors of the transferring country have another contextual background and thus different characteristics than actors of the receiving country. Second, the interactive process and the knowledge being transferred need to be embedded in the wider and structural context of the receiving country.

Methodology

This chapter builds upon qualitative case study research as its research methodology. By doing so, we aimed to describe and to understand the projects under study within their relevant contexts. This improved understanding is provided by analysing and comparing multiple projects that are embedded in a similar context (Yin, 2009). In this chapter, we compare three Romanian projects: the development of an Integrated Area Plan to create 'Room for the River' (case A); the implementation of a Flood Information and Warning System (case B); and the development of a Master Plan for Integrated Water Management (case C). All projects were implemented in collaboration between Dutch and Romanian actors in the period between 2008 and 2011 (see Figure 4.1 for the location of the case studies). In this period, the first author was based in Romania and collected data by means of direct observation, interviews and document analysis. She observed all major project activities (i.e. plenary meetings, workshops and field visits) and some of the project team meetings yet had no formal role in the design or implementation of the projects. To increase the reliability of her observations, she often cooperated with a Romanian co-observer and discussed the observations with project team members. She also regularly conducted informal interviews with project participants. After each project was completed, she conducted semi-structured interviews with the main project team members. The observational and interview data were complemented with written data that were retrieved, for example, from project reports, policy documents, websites and newspapers. For each

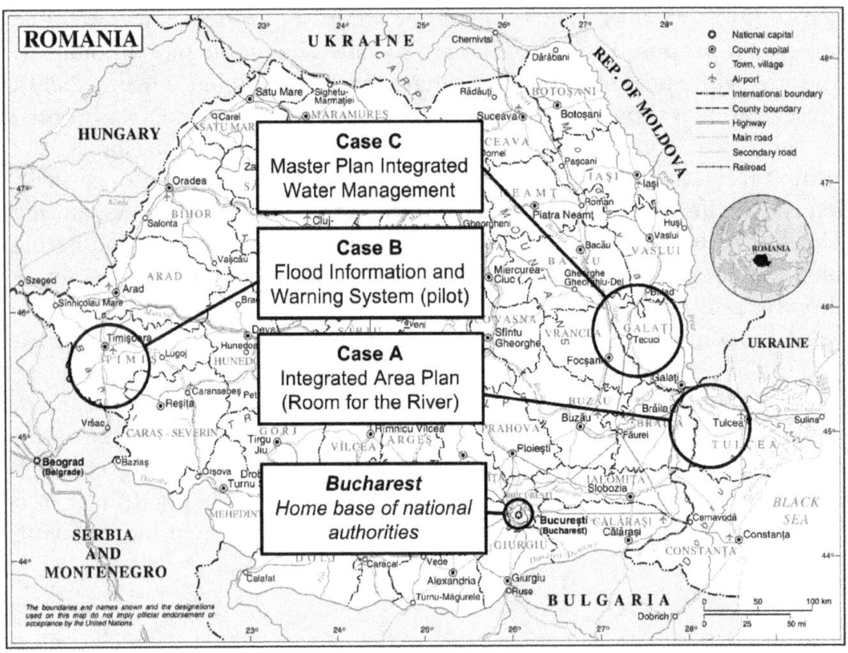

Figure 4.1 Map of Romania with the location of the three case studies (background map from United Nations, 2008).

project, a 'thick' description was prepared and published in the form of a case study report. These reports present detailed narratives of the project context, the interactive process and the characteristics of actors involved (Vinke-de Kruijf, 2011a, 2011b, 2012). The case studies were complemented with an analysis of Romania's wider and structural context. For this, the first author analysed relevant documents (including literature, newspaper articles, legislation and policies), participated in meetings, interviewed policymakers and experts and distributed a survey among operational flood risk managers.

4.3 Dutch-funded flood risk management projects in Romania: context and case studies

Dutch-Romanian cooperation setting

The Netherlands has a long history in water management. Within this context, it has long been active in supporting the solving of water management problems around the world. In its National Water Plan of 2009–2015, the Dutch government mentions two major reasons for supporting water projects around the world. First, it supports contributions to solving water-related problems (especially mitigation of climate change and realization of the Millennium Development Goal on water and environment). Second, it aims to strengthen the international position of the Dutch water sector by supporting collaborative projects with potential spin-offs in the form of follow-up projects. These objectives are especially realized through the exchange of knowledge and experiences (Min. V&W, 2009). It is within this context that the presented cases received the financial support of the Dutch government. Cases A and B were financially supported by Partners for Water, a programme that aims to stimulate the implementation of innovative projects of the Dutch water sector abroad. Case C received financial support from a foundation that supports international projects of Dutch water boards.

The collaboration between the Netherlands and Romania on water and environment dates back to 1995 when the Environmental Ministries of both countries signed a Memorandum of Understanding. The number of Dutch-funded projects increased substantially after 1998 when both countries signed a bilateral agreement. This agreement formed the basis for the implementation of various Dutch-funded projects in Romania. Between 1995 and 2007, about 140 projects in the field of environment and water (costing over 20 million euros) were implemented with financial support from programmes like the Social Transformation Programme Matra and the Emerging Markets Cooperation Programme PSO/PSOM.

In 2007, Romania became a member state of the European Union (EU). At this time, the Netherlands began to phase out its bilateral assistance. Although the number of Dutch programmes that support projects in

Romania decreased, various funding agencies continue to provide financial support. Furthermore, bilateral contacts between both countries continue to exist. In 2009, the Dutch Union of water boards and the National Administration for Romanian Waters signed a Memorandum of Understanding (RNE and EVD, 2009). Through this collaboration, the Dutch water boards intend to exchange knowledge with Romanian water authorities (Unie van Waterschappen, 2005). In 2008, the Netherlands Water Partnership established a Romania platform that aims to strengthen the position of the Dutch water sector in this country. This platform was established since Romania is expected to become an attractive market: huge investments are needed to bring environmental infrastructure in line with EU standards but domestic capacity and expertise is limited (Van Peppen, 2008). In the past years, the platform organized a wide range of activities to promote Dutch-Romanian cooperation in the water management domain. The platform also initiated in collaboration with the Union of Water Boards a Dutch-Romanian panel with senior executives and civil servants in 2009. This panel meets regularly to discuss various water management challenges that relate to living in a delta area (e.g. flood risk management, water quality and international river basin management). Completed and potential collaborative projects in the flood risk management domain, such as the presented cases, are also discussed in this panel.

Case A: 'Room for the River' and people in the Cat's Bend region

The objective of case A was to develop, in close cooperation with relevant stakeholders, an integrated plan that would contribute to flood risk reduction for the Cat's Bend region. This region is located just upstream of the Danube Delta where two major rivers (the Prut and the Siret) join the Danube. The project was initiated in 2006 and implemented between September 2008 and December 2009. The project design was inspired by the Dutch 'Room for the River' concept, which involves the integration of safety with other functions (e.g. nature and socio-economic development) and a preference for measures that create more space for the river rather than traditional measures, such as the further heightening of dikes. The project was also based on a Romanian study on the reassessment of the Danube floodplains. For the Cats' Bend region, this study recommended combining water storage with agriculture. The project involved a wide range of stakeholders including representatives of local and regional authorities, governmental and non-governmental organizations, and the private sector. This participatory process was supported by a team of two Dutch and four Romanian organizations. The team interviewed and invited stakeholders, prepared maps, visualized and conceptualized ideas, developed project materials, modelled the water system and calculated the impact of measures on the water level. The project resulted in three design concepts that could contribute to flood risk reduction along the Danube. The knowledge that was transferred and generated in the project was not used in any follow-up actions. The formal commissioner of the project, the Ministry of Environment, did not pay attention to the project results. Local and regional stakeholders wanted to implement some of the outcomes but were lacking necessary resources. As the project did not contribute to problem-solving or follow-up projects, it has not been effective from the perspective of the Dutch financers (Vinke-de Kruijf, 2011b).

Case B: implementation of FLIWAS in Banat region

Case B aimed to support Romanian water authorities through the pilot implementation of a Flood Information and Warning System (FLIWAS). FLIWAS is an internet-based application that helps to manage flood-related information before, during and after high water events. The application was recently developed in a European project and is currently being used by Dutch and German partners for the Rhine river basin. The pilot project in Romania was initiated in April 2009 and implemented between September 2009 and April 2010. The project was executed in cooperation between two Dutch consultants and four Romanian organizations. The pilot included the development of a Romanian FLIWAS environment (including the installation of an internet server and the training of potential users and administrators), the pilot implementation of FLIWAS at a regional water authority and various communication and dissemination activities. During the project, there were regular meetings with a wide range of actors and various work sessions at the regional water authority. Installing the FLIWAS on a national server constituted a major issue during the implementation process. Various experts had difficulties arriving at a mutual understanding. As a result, the server was installed much later than expected. To prevent further delay of the project, collected data were initially input on a Dutch server. The plan was to transfer these data to the Romanian server but it was unclear who was able and willing to do this transfer. In the end, data were never transferred and the regional authority did not begin using FLIWAS. The project therefore did not directly contribute to the reduction of flood risks. The project may still become effective as the Ministry of Environment was very interested in applying FLIWAS at a larger scale and also included the further development and implementation of a system like FLIWAS into a project proposal for the implementation of the European Strategy for the Danube river basin (Vinke-de Kruijf, 2011a).

Case C: integrated water management in the Tecucel river basin

The third case study concerns a project that was initiated by a Dutch regional water board following floods along the Tecucel River in 2007. The project aimed to develop an integrated framework (master plan) for the management of flood risks, drinking water and wastewater that would include several 'no-regret measures' (i.e. easily implemented measures that improve the water system). The project was implemented in cooperation between six Dutch organizations (including a water board, water company, municipality, governmental agency and two private companies) and five Romanian organizations (including three municipalities, a local water company and a local person of a regional water authority). Dutch ideas on interactive planning, which meant to enhance collaboration between various actors, provided the inspiration for the project. The project included visits by the Dutch team to Romania and also a visit each by the Romanian team and a regional water company to the Netherlands. While analysing the problems, potential solutions and ongoing policy developments, the project team realized that regional authorities already drafted master plans, one for drinking water and wastewater, and one for flood risk management. This led the teams to the conclusion that another master plan would be redundant. The Dutch team therefore decided to focus on the formulation and further development of two no-regret measures (i.e. a separate system for sanitary sewage and storm water runoff for the City of Tecuci and a flood retention reservoir along the Tecucel River). As these measures did not have any regional or national priority, the teams were not able to mobilize the necessary support and funds to further develop these measures. However, the project did form a basis for a formal agreement, a Water Partnership, between the Dutch water board and water company and the Romanian regional water company, which acquired the local water company during the project course. This partnership will form the basis for further exchange of knowledge and experiences with drinking water and wastewater management for the next three years. The project itself was not effective: it did not contribute to problem solving; neither did it generate any follow-up projects that would benefit the Dutch water sector (Vinke-de Kruijf, 2012).

4.4 Case study results: characteristics of actors involved

Motivations and the development of a joint motivating goal

In all cases, most actors were motivated by the international dimension of the project; they simply enjoyed participating in an international project. The opportunity to transfer knowledge was another common source of motivation. Dutch experts are formally interested both in sharing and acquiring knowledge, that is, learning from the experiences of other countries (Min. V&W, 2009). However, the cases show that Dutch experts were especially interested in sharing their knowledge: they believed that their knowledge could contribute to the solving of water-related problems. For some actors, knowledge sharing was important not only from a personal but also from their organizational perspective. This especially applies to the water board and the water company that participated in case C. Together with other water boards and water companies, they committed themselves to the realization of the Millennium Development Goals. Another motivation for sharing knowledge was that it provided experts with the opportunity to further develop and test their existing knowledge. An expert in case A mentioned that international projects are very useful training for landscape architects as they have to apply their skills in an unknown setting. Experts in case B were eager to test whether the FLIWAS technology was applicable in another setting. For the private organizations involved, the increased chance of being involved in other international projects played an important role. Experts mentioned that their company participated in the project to strengthen their international network, position and portfolio.

A major motivation of the Romanian actors was to reduce water-related problems in the region under concern. Most of them highly valued Dutch knowledge and believed that the project could contribute to solving certain issues. They were therefore interested in acquiring Dutch knowledge concerning the application of a certain method, concept or technology or about potential solutions to concrete problems. They further expected that the international dimension of the project would help to raise national attention for local issues (local actors in case C) or would make them known as being forerunners (regional authority in case B). At the same time, actors in case A initially doubted whether their ideas would form the basis for the project outcomes and whether these outcomes would influence the decision-making process. Some of them were also sceptical about the role of Dutch experts and the added value of their expertise. When they witnessed that their ideas were taken into account, they also started to support the implementation of the project objectives. Like the regional and/or local actors in the other case studies, they eventually also saw the project as an opportunity to influence decision making. However, despite the objectives of all cases being (eventually) strongly supported by local and/or regional actors, they were often not implemented. One of the main issues was that the project goals were lacking support at the national level. Cases A and B were both initiated, and thus initially supported, by actors at the national level. In case A, this support diminished especially after the State Secretary was replaced. The Ministry of the Environment eventually ignored the results and the request of local authorities to support them with the implementation of results. In case B, the ministry stayed involved and was also motivated to include an application like FLIWAS in a new project proposal. Case C was designed in a bottom-up fashion where the ministry was asked to have a consultative role. Although local and regional actors were keen to involve the ministry, they were never able to raise sufficient support to further elaborate the project results.

In all cases, the motivations of some of the actors involved also changed during the course of the interactive process. Although Dutch experts in all cases remained committed to finalizing the project, several experts became less motivated to continue the collaboration. Experts of a Dutch governmental agency (involved in cases A and C) explained that their organization was formally committed to international water management but that they discovered that the actual organizational support for such projects was limited. This lack of support is among the reasons why they became less motivated to continue working on such projects. The collaboration with Romanian partners was also often not as expected. In case A, several experts were disappointed about the expertise that was contributed by one of the Romanian partners and therefore reluctant to collaborate with them in future projects. In cases A and C, experts had doubts whether some of their Romanian partners were actually willing to make an effort to solve their water problems. This is why experts in case A left the initiative for follow-up with the Romanian actors.

Cognitions and the creation of 'negotiated knowledge'

All cases were based on the recognition that the frequency and intensity of floods are increasing and that the reduction of flood risks and related issues forms a pressing issue. This cognition formed the main input for the project proposals that were prepared by Dutch experts following discussions with Romanian actors but this cognition often changed during the interactive process. In case A especially, the initial framing of the project by Dutch experts did not correspond with the actual problems in the area. Dutch experts expected that floods were a major problem in the region. However, when the project started, it soon appeared that Romanian partners and stakeholders did not share this cognition. In the opinion of the Romanians, the main problems in the area were of a different nature and included unequal access to resources, drought, declining biodiversity and a lack of public participation. Dutch experts therefore decided to broaden the project scope. The developed plans were still largely as initially expected by the Dutch experts involved. They created more 'Room for the River' and therefore contributed to flood risk reduction but also to regional economic development and a better micro-climate. In case B, floods were also perceived as being an important issue by the Romanian partners. In this case, some of the Dutch experts also discovered – although to a lesser extent – that the actual problems were different from expectations. A Dutch expert explained, for example, that he only realized during the project that FLIWAS could really reduce the vulnerability of the emergency management system that resulted from an overdependence on individuals and individual knowledge. Case C was also initiated in order to reduce flood risks but the scope of the project was broadened already from the beginning to also include problems in the water services sector. Only after the project started, the Dutch experts discovered that relevant actors did not know each other yet. In the opinion of Dutch experts, one of the main project results was therefore that the project created a connection between these actors.

All three cases concerned the transfer of 'Dutch ideas' to Romania. As these ideas had not been implemented in Romania before, both Dutch and Romanian actors were uncertain whether these ideas could be successfully applied in the Romanian context. While case A was not very effective, most of the actors involved were very positive about the applicability of the concept and method used. Case B proved to Dutch experts that FLIWAS could be implemented in Romania. Romanian experts only partly shared this cognition. They doubted whether it could be implemented in other parts of Romania and – as they never began to use FLIWAS – also doubted its actual usability. The integrated and bottom-up approach in case C appeared to be rather ineffective. Dutch experts concluded that the integrated approach was of added value but it should be reconsidered and that a better balance was needed between bottom-up and top-down. However, several Romanian experts had doubts about whether such an approach could ever be effective in Romania.

Resources and the mobilization of necessary resources

In all cases, Dutch actors provided most of the financial resources for project implementation. The costs of the projects were partly covered by a Dutch funding agency (about 80 per cent in the cases A and B and about 50 per cent in case C) while the remaining costs were paid for by the Dutch organizations. Human resources were contributed both by the Dutch and Romanian organizations involved. In case A, Dutch organizations paid for the involvement of Romanian partners and covered other costs (such as meals and meeting locations). In cases B and C, Dutch experts collaborated more closely with Romanian authorities who could often arrange such resources for free. One of the critical issues in all cases was the financing of follow-up steps. All cases predominantly involved local and regional actors (e.g. representatives of governmental authorities, non-governmental organizations or other stakeholders). However, follow-up actions for the cases usually required considerable investments that were beyond the capacity of these actors. In cases A and C, actors from the area were motivated to continue with the project results but lacked the capacity to mobilize the resources for this. In case B, the ministry was more closely involved and was in the position to integrate a tool such as FLIWAS into a proposal for external funds.

Since all cases concerned the transfer of a 'Dutch idea' (e.g. 'Room for the River', 'Sketch Match', FLIWAS or integrated water management) from the Netherlands to Romania, Dutch actors were important sources of knowledge. Projects were usually driven by general knowledge and initiatives of the Dutch experts involved as Romanian actors lacked the expertise to apply these ideas. In this sense, the knowledge transfers took place in a typical development setting in which local knowledge meets general knowledge of external experts (Rist *et al.*, 2006). In all cases, a Dutch expert was responsible for the overall project management. In addition, Dutch experts often also made an analysis of which actors or stakeholders should be involved as Romanian actors were not familiar with the key concept, method or technology being transferred. Especially in case B, Dutch experts could also contribute some context-specific knowledge as they had been involved in Dutch–Romanian projects before. Romanian actors were important sources of context-specific knowledge and often responsible for the organizational aspects, such as inviting stakeholders and arranging meeting locations. To manage the involvement of Romanian actors, each project had a Romanian project coordinator (and a Romanian project secretary in cases B and C). In case A, experts had a facilitative role and the results were mostly based on context-specific knowledge that was provided by stakeholders. Also in the other cases, there was attention to the inclusion of stakeholder knowledge but, in practice, professional experts provided the most knowledge.

Romanian experts were generally valued for their high level of technical knowledge but their knowledge contribution was on several occasions less than expected by Dutch experts. In case A, Dutch experts expected that the participating institute would really contribute to the project content (e.g. hydraulic modelling, processing of geographic information and preparing area maps). Only during the project did it appear that the institute could not provide such expertise. This was one of the reasons why the project outcomes were slightly different from expectations. It also caused some experts to have reservations about working with the same institute in any future project. Experts in case B had similar experiences. In this case, the installation of the server was expected to be rather straightforward. Only during the project did it appear that the appointed expert did not have specific experiences in this domain. In case C, one of the main misunderstandings between Dutch and Romanian experts was the need and availability of a master plan. Dutch experts only realized halfway through the project that a master plan was redundant as two regional master plans had just been approved. This shows that in all cases, Dutch experts attributed more knowledge to Romanian experts than they could contribute.

4.5 Analysis and discussion: on the role of contextual factors

The influence of differing wider contexts

The previous section shows that in all cases there were major differences between Dutch versus Romanian actors in terms of motivations, cognitions and resources. In general, Dutch experts wanted to share knowledge and provided most knowledge and financial resources. Romanian actors participated in order to learn and acquire knowledge and to gain (financial and organizational) assistance with problem solving. The country-specific background of both actor groups explains these differences. In terms of problem context, the Netherlands is a densely populated country with low-lying polders in which floods are extremely dangerous. In practice, floods rarely occur as the country adopted very high safety standards (Van Alphen and Lodder, 2006). In Romania, floods have become a yearly recurring issue during the last decades. They are not only causing major economic damages but also caused an average of thirteen casualties per year between 1969 and 2006 (GoR, 2010). As the intensity and frequency of floods are on the increase, Romanian actors are eager to acquire fresh knowledge about how to solve these problems. Moreover, Romanian actors perceive the Netherlands as one of the most – if not the most – advanced countries in flood risk management.

In the last decades, both countries also went through very different political and economic developments. Shortly after the Second World War, the Netherlands became closely involved in the establishment of the EU whereas Romania was ruled by a communist regime. Since the 1990s, Romania engaged in a transition process that aims to bring the country's political and economic system in line with mainstream European systems (Gallagher, 2005). In economic terms, the gross domestic product of the Netherlands was among the five highest and Romania among the two lowest of all EU member states in the past years (Eurostat, 2012). Being a relatively rich and export-oriented country, international assistance has always had a rather high priority in the Netherlands. This has created the opportunity to also pay for the transfer of knowledge to less developed countries. This financial support is attractive for Romania, which depends heavily on external funds for the implementation of any project. To gain access to funds is especially challenging for local and regional actors who are therefore eager to participate in any project. What also plays a role in this is Romania's wish to 'catch up' with other EU member states (Rose, 1993). In case C, for example, actors also perceived the project as a means to learn from another EU member how to apply for EU funds.

As both countries need to comply with the same EU rules and legislation, their structural context for water management is partly similar. However, there are also differences that form an incentive for knowledge transfer. Especially in case A, actors were very eager to learn about the application of the 'Room for the River' concept or the interactive design method as tools for integrated and participatory water management. The Netherlands is known for developing these ideas and has experimented with them for years (van Ast, 1999; Mostert, 2006), while such approaches are still very new in Romania. Non-governmental organizations, especially, were very interested in such approaches, which are required by various European Directives (including the Flood Directive) but often poorly implemented (Teodosiu, 2007; Teodosiu *et al.*, 2003). Dutch actors perceive the sharing of their knowledge in these domains also as one of their tasks as a more experienced country.

Literature often emphasizes that contextual differences complicate policy transfers (Mossberger and Wolman, 2003; Swainson and de Loe, 2011). However, the case of the Netherlands and Romania shows that such differences are also an incentive for knowledge transfer and therefore an important source of motivation. However, the cases also show that contextual differences can challenge the effectiveness of such transfers as it implies that actors have different cognitions and resources. In case A, Dutch actors framed flood risks as an urgent problem that was rooted in a lack of space for the river. This cognition was not shared by many of the Romanian actors involved. Only after adjusting the project scope could they arrive at a common goal. The differing problem contexts of both countries provide an explanation for these differing cognitions. While this difference did not affect the effectiveness of this project, differences in terms of structural context were influential. In the Netherlands, 'Room for the River' projects are part of a national programme (Min. V&W, 2006). In Romania, such a programme did not exist and the measures were not sufficiently supported by national actors. As a result, it was not possible to mobilize financial resources for implementation. The effectiveness of case B was especially affected by differences in terms of knowledge resources, which led to problems with the installation of the server and had a negative impact on the effectiveness of the case. The underlying problem was that Dutch experts expected Romanian experts to contribute a slightly different type of technological knowledge than Romanian experts expected to be necessary. In case C, Dutch actors expected – on the basis of their own structural context – that Romanian actors were used to collaborating with each other. Later it appeared that in the Romanian structural context, there was very little interaction between actors with a role in water management. These differences explain why actors understood the meaning of existing master plans only when the project was already halfway complete.

Embedding in the dynamic Romanian context

Analysis of the cases shows that projects that involve the international transfer of knowledge are often less effective than planned. None of the projects contributed directly to the reduction of flood risks and only case B is likely to create follow-up projects for the Dutch water sector. One of the main explanations for these differences is that in cases A and C, there were no powerful players 'pulling-in'. In other words, resourceful actors were not convinced about the need to implement or to elaborate the project results. In case A, a powerful actor (the State Secretary for Water) was involved in project preparations and supported the project idea. However, this national support faded when – following elections – this person was replaced. Also at the regional level, several key actors were replaced following elections. These political changes would not have had such a negative impact on the project if the 'Room for the River' concept had been widely supported. This was not the case since the concept was rather controversial, which made it difficult to raise support for the project among executives and politicians. This was even more so as the project was finalized in the middle of another election period. A mismatch between the project results and the political context thus explains why resourceful actors would not support the project. What is remarkable is that the process was initially also not supported by regional/local actors. This related to a combination of contextual factors, such as distrust in the government, negative experiences with participation and high dependence on the national government for resources. This negative motivation eventually altered and therefore had no negative impact on the projects' effectiveness.

Case B was implemented in a relatively short period, which is one of the reasons that it was not affected by any changes in the political context. What contributed to the effectiveness of this case was that powerful actors (at the national level) supported the project from the beginning until the end. Improvement of information and warning systems was one of their key priorities and they therefore deliberately looked for ways to integrate a tool like FLIWAS in one of their project proposals. This confirms that transfers are likely to be supported by policymakers – and thus more likely to be implemented – when they are consistent with political consensus (Rose, 1993). In this case, it was also less difficult than in the other cases to embed the project as the project involved several Dutch experts who were familiar with the Romanian context and policy network due to their extensive project experience in the country. However, whether the tool will eventually be used is still to be seen. Experts explained that its actual use may become problematic, among others, because automated measurement stations are not functioning well and end users lack access to modern information and communication means. In other words, they doubt whether the tool fits Romania's technological context.

In case C, the entire project developed very differently than planned. One of the issues during the process was that none of the actors had an adequate insight in relevant policy processes. One of the underlying problems was that there is very little interaction between various Romanian actors with a role in water management. This confirms that incoherent governance may indeed affect project implementation as it leads, among others, to more uncertainty and a failing capacity to recognize win-win situations (De Boer and Bressers, 2011). When information became available, the initial project objective – a master plan – became redundant. However, there were also no project resources to develop a completely different project. Besides this, the project teams also failed to mobilize financial resources for the implementation of their common goal. This failure closely relates to the misfit between the Romanian situation and the integrated and bottom-up project approach. An integrated approach was difficult to realize since European and national funds are not organized in this way. The adoption of a bottom-up approach was also not effective. After the communist period, a decentralization process begun but developed in a rather inconsistent manner. While competences were transferred to a lower level, associated power and finances were not. The decentralization process also did not account for the lack of expertise and training of local officials (Bădescu *et al.*, 2004). The funding programmes that were of interest in case C were only accessible for regional and national authorities. These actors had not been involved in the project and did not give priority to the problems in the area. What also did not help the mobilization of resources in cases A and C was that national funds decreased considerably as a result of the global economic crisis.

4.6 Conclusions

The aim of this chapter is to provide insight into how contextual factors influence the effectiveness of projects that involve an international knowledge transfer. We conceptualize these projects as actor-interaction processes that are embedded in wider, structural and specific contexts. The Contextual Interaction Theory shows that these contexts only exert an influence on such processes – and their effectiveness – via their influence on the motivations, cognitions and resources of actors involved. To better understand the influence of specific conditions and circumstances, we analysed three Dutch-funded projects that involved the transfer of knowledge about flood risk management from the Netherlands to Romania. Due to contextual differences, there were major differences between Dutch and Romanian actors in terms of motivations, cognitions and resources. These differences provide an incentive to collaborate and therefore an important source of motivation. They also imply that actors have different cognitions and resources, which can result in disagreement and misunderstanding. A further analysis of the projects' embedding into the Romanian context highlights that the effectiveness of the projects was correlated with the involvement and support of national actors. In other words, the cases confirm the importance of powerful players 'pulling-in' (Kroesen *et al.*, 2007). Case C in particular further reveals the fragmented governance setting of Romania as another factor that inhibited a successful knowledge transfer. This finding confirms the notion that fragmented governance may seriously hamper policy implementation and thus effective knowledge transfers (De Boer and Bressers, 2011).

References

APFM (Associated Programme on Flood Management) (2012) website, www.apfm.info, accessed 27 April 2012.

Bădescu, G., Sum, P. and Uslaner, E.M. (2004) 'Civil society development and democratic values in Romania and Moldova', *East European Politics and Societies*, vol. 18, no. 2, pp. 316–241.

Bressers, H.T.A. and Kuks, S.M.M. (eds) (2004) *Integrated Governance and Water Basin Management: Conditions for Regime Change towards Sustainability*, Kluwer Academic Publishers, Dordrecht/Boston, MA/London.

De Boer, C. and Bressers, H. (2011) *Complex and Dynamic Implementation Processes: The Renaturalization of the Dutch Regge River*, University of Twente, in collaboration with the Dutch Water Governance Centre, Enschede.

EEA (European Environmental Agency) (2008) 'Impacts of Europe's changing climate: 2008 indicator-based assessment', EEA Report No. 4/2008.

Eurostat (2012) 'GDP per capita in PPS', online, available at: www.epp.eurostat.ec.europa.eu/tgm/table.do?tab=table&init=1&plugin=1&language=en&pcode=tec00114, accessed 22 June 2012.

Gallagher, T. (2005) *Theft of a Nation: Romania since Communism*, C. Hurst and Co., London.

GoR (Government of Romania) (2010) 'Strategia Nationala de Management al riscului la inundatii pe termen mediu si lung (National Strategy for Flood Risk Management for the medium and long term) – Annex to Governmental Decision No. 846/2010', *Monitorul Oficial*, vol. 626, no. 6, September 2010.

Kroesen, O., De Jong, M. and Waaub, J.P. (2007) 'Cross-national transfer of policy models to developing countries: Epilogue', *Knowledge, Technology & Policy*, vol. 19, no. 4, pp. 137–142.

Min. V&W ([Dutch] Ministry of Transport, Public Works and Water Management) (2006) 'Spatial Planning Key Decision Room for the River: Approved Decision 16 December 2006', Ministry of Transport Public Works and Water Management, Den Haag, online, available at: www.ruimtevoorderivier.nl/media/21963/pkb%204%20approved%20decision%20h01-h086.pdf, accessed 19 November 2012.

Min. V&W ([Dutch] Ministry of Transport, Public Works and Water Management) (2009) 'National Water Plan 2009–2015', Ministry of Transport Public Works and Water Management, Den Haag.

Mossberger, K. and Wolman, H. (2003) 'Policy transfer as a form of prospective policy evaluation: Challenges and recommendations', *Public Administration Review*, vol. 63, no. 4, pp. 428–440.

Mostert, E. (2006) 'Integrated water resources management in The Netherlands: How concepts function', *Journal of Contemporary Water Research and Education*, vol. 135, no. 1, pp. 19–27.

Mostert, E., Pahl-Wostl, C., Rees, Y., Searle, B., Tàbara, D. and Tippett, J. (2007) 'Social learning in European river-basin management: Barriers and fostering mechanisms from 10 river basins', *Ecology and Society*, vol. 12, no. 1, art.19.

Unie van Waterschappen (2005) 'Cooperation between Apele Romane and Dutch Water Boards', Report of visit to the Netherlands, 23–28 October 2005, Unie van Waterschappen, Den Haag.

Owens, K.A. (2008) 'Understanding how actors influence policy implementation: A comparative study of wetland restorations in New Jersey, Oregon, The Netherlands and Finland', PhD thesis, University of Twente at Enschede.

Rist, S., Chiddambaranathan, M., Escobar, C. and Wiesmann, U. (2006) '"It was hard to come to mutual understanding...": The multidimensionality of social learning processes concerned with sustainable natural resource use in India, Africa and Latin America', *Systemic Practice and Action Research*, vol. 19, no. 3, pp. 219–237.

RNE and EVD (Royal Netherlands Embassy and Agency for International Business and Cooperation) (2009) 'Environment and Water Projects in Romania', 3rd edition, April 2009, Royal Netherlands Embassy and Agency for International Business and Cooperation, Bucharest, Den Haag.

Rose, R. (1993) *Lesson-Drawing in Public Policy: A Guide to Learning across Space and Time*, Chatham House Publishers, Chatham, NJ.

Samuels, P., Klijn, F. and Dijkman, J. (2006) 'An analysis of the current practice of policies on river flood risk management in different countries', *Irrigation and Drainage*, vol. 55, no. S1, pp. 141–150.

Sehring, J. (2009) 'Path dependencies and institutional bricolage in post-Soviet water governance', *Water Alternatives*, vol. 2, no. 1, pp. 61–81.

Stone, D. (1999) 'Learning lessons and transferring policy across time, space and disciplines', *Politics*, vol. 19, no. 1, pp. 51–59.

Swainson, R. and de Loe, R.C. (2011) 'The importance of context in relation to policy transfer: A case study of environmental water allocation in Australia', *Environmental Policy and Governance*, vol. 21, no. 1, pp. 58–69.

Teodosiu, C. (2007) 'Challenges for integrated water resources management in Romania', *Environmental Engineering and Management Journal*, vol. 6, no. 5, pp. 363–374.

Teodosiu, C., Barjoveanu, G. and Teleman, D. (2003) 'Sustainable water resources management 1. River basin management and the EC Water Framework Directive', *Environmental Engineering and Management Journal*, vol. 2, no. 4, pp. 377–394.

United Nations (2008) *Map No. 3661 Rev. 5*. Department of Field Support, Cartographic Section.

Van Alphen, J. and Lodder, Q. (2006) 'Integrated flood management: Experiences of 13 countries with their implementation and day-to-day management', *Irrigation and Drainage*, vol. 55, no. S1, pp.S159– S171.

van Ast, J.A. (1999) 'Trends towards interactive water management: Developments in international river basin management', *Physics and Chemistry of the Earth, Part B*, vol. 24, no. 6, pp. 597–602.

Van Peppen, D. (2008) 'Concept strategie NWP Roemenie Platform [Draft strategy NWP Romania Platform]', Netherlands Water Platform (NWP), Den Haag.

Vinke-de Kruijf, J. (2011a) 'The role of Dutch expertise in Romanian water projects: Case study "Pilot implementation FLIWAS in Banat region, Romania" ', CE&M research report 2011R-02/WEM-002, University of Twente, Enschede.

Vinke-de Kruijf, J. (2011b) 'The role of Dutch expertise in Romanian water projects: Case study "Room for the River in Cat's bend, Romania" ', CE&M research report 2011R-001/WEM-001, University of Twente, Enschede.

Vinke-de Kruijf, J. (2012) 'The role of Dutch expertise in Romanian water projects: Case study "Integrated Water Management in the Tecucel River Basin" ', CE&M research report 2012R-001/WEM-001, ISSN 1568-4652, University of Twente, Enschede.

Vinke-de Kruijf, J., Augustijn, D.C.M. and Bressers, J.T.A. (2012) 'Evaluation of policy transfer interventions: lessons from a Dutch–Romanian planning project', *Journal of Environmental Policy and Planning*, vol. 14, no. 2, pp. 139–160.

Vinke-de Kruijf, J., Hulscher, S.J.M.H. and Bressers, J.T.A. (2011) 'Knowledge transfer in international cooperation projects: experiences from a Dutch–Romanian project', Conference Paper, 5th International Conference on Flood Risk Management (ICFM5), 27–29 September, Tokyo.

WMO (World Meteorological Organization) (2009) 'Integrated flood management: Concept paper', WMO-No. 1047, World Meteorological Organization, Geneva.
Yin, R.K. (2009) *Case Study Research: Design and Methods*, Sage Publications, Thousand Oaks, CA.

5 Testing of the Contextual Interaction Theory in the evaluation of cooperation and collaboration of water management projects in India

Katharine Owens, Marcia Hughes and Ellen Skoczenski

5.1 Introduction

It is common practice for volunteers from industrialized nations to visit developing countries in efforts to transfer knowledge and expertise on a range of issues (Engineers without Borders, 2012). This chapter evaluates a University of Hartford (Connecticut, USA) student–faculty team implementing a water management project in India. This team brings know-how and expertise to a rural Indian community to solve a public problem, namely water drainage on a public street. The chapter asks whether when researchers or volunteers visit developing nations to transfer policy, are they truly collaborating with local populations? The Contextual Interaction Theory (CIT) is used to analyse project implementation, comparing the theoretical prediction to the actual outcome. The theory is also used as a tool to ask whether the theory-predicted cooperative interaction meets key characteristics of interpersonal collaboration (a shared purpose, dialogue, decision making, action and evaluation). Analysis focuses on how the theory in conjunction with a collaboration assessment can highlight whether a policy transfer project, designed to be a cooperative effort, is truly collaborative.

Citizens of the developed world often feel compelled to embark upon projects in developing countries to transfer knowledge and expertise. These actors seek to develop and implement programmes to address perceived problems. It is possible that intention and what happens 'on the ground' can be quite different, both in terms of knowledge transfer toward implementation and interpersonal collaboration. Similarly, it does not follow that putting a programme in place yields the desired results. This analysis of one case of policy transfer evaluates both the implementation and the level of collaboration among stakeholders. Researchers often focus on the implementation stage of the policy process as a point where intentions and outcomes diverge (Pressman and Wildavsky, 1973; Palumbo and Calista, 1990; Hill and Hupe, 2002; Bressers,

2004). Definitions of implementation emphasize different concepts, ranging from those who see implementation as the realization of a task, to those who envision it as interactions that may include alterations to, or even the blocking of, a policy (Wittrock, 1985; Palumbo and Calista, 1990; O'Toole, 2000). Implementation is defined as where the intentions of a policy or programme are put into action in real life. This stage in the process is evaluated to better understand participants' roles, any impediments to change, and the effectiveness of the policy transfer, which are explored through the lens of collaboration (see Chapter 2 for a thorough treatment of policy transfer). If policy transfer can take on a tone of coercion, it is important that researchers explore the nature of collaboration (Dolowitz and Marsh, 1996).

This chapter describes use of the CIT (a newer version of the theory than the one used here is elaborated in Chapter 3) as an analysis tool to systematically assess the implementation of this knowledge transfer and to further delve into an evaluation of collaboration. This is not necessarily the implementation of stated Integrated Water Resources Management (IWRM) goals, it is instead part of an ongoing relationship between the Engineers without Borders (EWB), Navjyoti India Foundation, and villagers of Abheypur, India that seeks to improve livelihoods by transferring knowledge around water systems from engineers and engineering students in a developing country context. So, while the student–faculty team does not explicitly state IWRM goals, the underlying concepts are implicit in a great deal of the work the team does.

5.2 Collaboration in implementation

Several researchers promote collaboration as an ideal. Gajda and Koliba write that collaboration is necessary to address 'complex societal issues and successfully reach ... essential organizational outcomes' (2007, p. 26). In managing water resources in particular, Pahl-Wostl *et al.* (2007) find that a paradigm shift is needed, one that recognizes the importance of stakeholder involvement and collaboration. For many who promote collaboration, the goal is to develop a community of practice: 'the embodiment of interpersonal collaboration within an organization in which the individual members of a social learning system share common practices and work together to achieve mutually desired outcomes' (Gajda and Koliba, 2007, pp. 26–27). Preskill (2009) writes that 'environmental evaluations are about values and that *stakeholders must be involved in the design and implementation of an evaluation* as well as be identified as the users of the evaluation findings' (p. 99, our emphasis). Pushing this concept even further, David Fetterman (2001) envisions a future where evaluation takes the tone of fully collaborative processes. He writes, 'Citizens will come to the table with a basic knowledge of evaluation as part of their democratic literacy skills' (p. 381). He advocates training clients to evaluate themselves, in what he calls empowerment evaluation (Fetterman, 2002). According to Fetterman (2002), empowerment evaluation uses the participant's own perspective, 'help[ing] people help themselves and improve their programs using a form of self-evaluation and reflection' (pp. 89–90). Although project partners in this case are not currently engaged in Fetterman's empowerment evaluation, Fetterman's work shows the range of comprehensive collaboration being pursued by other researchers, who go far beyond the level of collaboration described in this project.

While evaluators or implementers may be in favour of collaboration, this does not mean it is necessarily forthcoming. Gajda (2004, p. 66) writes of the difficulty in grasping the concept,

> Its definition is somewhat elusive, inconsistent, and theoretical. In its overuse, the term 'collaboration' has become a catchall to signify just about any type of interorganizational or interpersonal relationship, making it difficult for those seeking to collaborate to put into practice or evaluate with certainty. In my experience as a program evaluator, most practitioners working to form strategic alliances are not sure about what collaboration looks and feels like. They are not sure if their collective actions constitute true collaboration and they do not know how to determine if the structural, procedural, and interpersonal relationships between the partners are as healthy as they could be.

Certainly programme evaluation and implementation are complex and difficult tasks; using the implementation process as a way to reflectively collaborate makes the process even more so. Though a thorny issue, the characterization of this transfer process as collaborative is a goal of this work. That being said, reflective researchers must question: would researchers realize it if their work was not collaborative? Should researchers trust their own conceptions or evaluate the process to confirm or deny collaboration? In later research, Gajda and Koliba (2007) describe key characteristics of interpersonal collaboration to be a shared purpose coupled with a 'cycle of inquiry' composed of dialogue, decision making, action, and evaluation (p. 29). This assessment uses their criteria to evaluate collaboration.

Collaborative processes have been under-operationalized in the field of evaluation, and there are a lack of studies that explore the relationship between the quality of collaboration and project outcomes. The analysis in this paper aims to elucidate this relationship. This chapter assesses the implementation of one water project in India, coupling analysis using the CIT with Gajda and Koliba's criteria of collaboration. These tools are used in conjunction to assess the implementation process and the collaboration of American and Indian stakeholders. The application of the CIT is used to evaluate how stakeholder motivation, information, and power influence the decision-making process. Gajda and Koliba's collaboration criteria are applied to evaluate the character of this transfer process.

5.3 The Contextual Interaction Theory

The CIT analyses key actor characteristics to shed light on *how* actors implement. Chapter 3 describes the most current elaboration of the theory. This analysis employs an older model of the theory, which was applied in Owens (2008) and is based most closely on the elaboration in Bressers (2004). Using independent variables of motivation, knowledge and power, the theory provides a hypothesis predicting the interaction that will occur among process participants. The variables motivation (Gross *et al.*, 1971; Ball, 1976; Nakamura and Smallwood, 1980; Larson, 1980), knowledge (Gross *et al.*, 1971; Levitt, 1980; Edwards, 1980; Williams, 1980, 1982) and power (Bunker, 1972; Berman, 1980; Raelin, 1980, 1982; Davies and Mason, 1982, Ackermann and Steinmann, 1982; Browne and Wildavsky, 1984) appear often in the implementation literature. Analysis focuses on implementation, simplifying an implementation situation into the assessment of two actors, the implementer and the target. The implementer is the group putting the changes into place, while the target is the group impacted by these changes. Though this is a two-actor model, it can account for other actors in an implementation process. These actors can be divided along meaningful lines into those supporting either a target or implementer perspective for a given issue, though perhaps in informal groups or coalitions. This assessment is an evaluation of the motivation, knowledge and power of these actors. Based on the values of these characteristics, the CIT creates a hypothesis about the kind of interaction process that will occur. This prediction can take one of many forms, including active cooperation, passive cooperation, forced cooperation, opposition, obstruction, joint learning or no interaction (see Bressers, 2004).

5.4 Principles and theory of collaboration

Any collaborative process involves two or more people coming together with a shared purpose and engaging in a 'cycle of inquiry' that includes dialogue, decision making, action and evaluation (for a full description see Gajda, 2004; Gajda and Koliba, 2007). Collaborative efforts are increasingly seen as the primary method for achieving goals. Importantly, intra- and interorganizational efforts are dependent upon 'interpersonal relationships': the ways in which individual participants personally connect with the social purpose of the collaboration and with each other. In any collaborative process, the partners need to have a shared understanding of what they hope to achieve, they have to have confidence in the collaborative process itself, and everyone has to participate in making decisions about their work together (i.e. shaping and implementing strategies). Effective collaboration therefore requires trust which needs time and energy to develop. Consistent representation at meetings (or other forums), cross-participation in related initiatives, and agreement about goals and strategies are all important for developing a successful partnership. In addition, the collaborative group needs to allow and perhaps even be prepared for the interpersonal conflict that is a normal part of the process as organizational efforts become more integrated and personal involvement increases (Gajda, 2004; Gajda and Koliba, 2007). Rather than hierarchical relationships, a community of practice should capitalize on interpersonal collaboration. The CIT is used to test whether the motivation, knowledge and power levels of implementers and targets will produce a cooperative interaction.

5.5 Project description

Since 2007, the University of Hartford and the Hartford Professional Chapter of EWB have been working in collaboration with villagers and members of Navjyoti India Foundation to make sustainable improvements in water supply and sanitation in Abheypur, a rural village located in Sohna Block, Haryana State, India. The team has developed and implemented one project per year for the past four years. Because of these efforts, families who previously depended on other communities and households for their water supply now have consistent access to water close to their home, and the women and girls who collect the water are no longer walking long distances.

The project under evaluation here involved implementing a soak pit on a village street in 2010. A hard-packed earthen street inhibits drainage. When rains occur, particularly heavy seasonal rains, the water does not soak into the substrate but instead forms runoff and small pools of unclean standing water. This provides an area for insect breeding, inhibits traffic along a roadway, and is unsightly. A soak pit is a water management tool installed on an earthen road to improve drainage. The installation of a soak pit involves digging a large hole in such a road, and then filling the hole with a substrate material, such as sand, that can allow the water to soak down and slowly become absorbed into the surface, therefore improving its consistency.

The project was developed through discussions with the Navjyoti India Foundation and local villagers in Abheypur, India. On earlier trips student–faculty team members observed water flooding and pooling on the street. In surveys, local villagers mentioned concerns about this issue. The student–faculty team then addressed this public issue by implementing the knowledge transfer of a soak pit.

Setting

When one considers the dynamics of the India project (e.g. student–faculty team, language, caste, gender, money and time) it is clear that there is a great deal of room for misinterpretation, confusion and miscommunication. The student–faculty teams include both engineers and social scientists, therefore participants implementing a technical project as well as those evaluating social phenomena. These teams are for the most part American and English-speaking, though occasionally include Indian-Americans. When in India, the team works with the Indian partner organization (Navjyoti) and through language volunteers (Evaluserve) to collect data, interact, manage projects and get input from local villagers. Despite this language assistance, working across languages adds a layer of complexity to the project. While caste is no longer official, small inclusive groups remain within the village that have few interactions with members of other groups, though villagers may explain this as class not caste. In addition, the student–faculty teams include women in positions of leadership, which is atypical in a rural village where gender strongly influences societal roles. While student–faculty team members may consider themselves middle-class, comparatively they are perceived in this village as fairly wealthy. Finally, the university trips take place during a short winter term, meaning the team spends between ten days and two weeks in India. This is a small amount of time to build trust among partners, implement new projects, and manage and evaluate previous projects. When implementing projects in Abheypur, the student–faculty team must not only implement, but must also reflectively manage projects, and work within these many constraints. While these are serious concerns, they are no different from many of the issues encountered by any group working across nationalities, cultures and language on sustainability (or other) projects. In this case, a long-term commitment as well as working with local stakeholders (i.e. Navjyoti India Foundation) helps to address some of these concerns.

Stakeholders

A cross-discipline of University of Hartford students and faculty, and professional engineers working through the EWB humanitarian organization provide consultation and expertise in assessment and design of this water-related project. EWB also focuses on capacity building and sustainability, and developing a replicable model for transferring knowledge to others.

Navjyoti India Foundation has a long history of working in rural development (~20 years). Their programmes emphasize strong advocacy and connection with community. As such, Navjyoti India Foundation is a critical partner in facilitating learning about the community's perspective and existing infrastructures, and ensuring that there is community participation, and ownership and maintenance of projects over time.

The village leaders and other community stakeholders make final decisions on selection of projects, review designs, and assist in selection of local materials. During implementation, their role is to organize the work force and lead the work. Budget needs and funding sources are determined for each project and shared by the three above-noted partners.

Other critical partners include Evaluserve Inc., a local corporation that provides volunteers to serve as translators and logistical support for ground transportation and purchase of materials, and Pathways World School, a private boarding school that provides the student–faculty team members with free room and board.

In theoretical parlance, the 'implementer' is the student–faculty team. Other groups supporting this perspective (members of this loose coalition) would be the Indian partner Navjyoti, and the professional EWB chapter (who also work in Abheypur, but were not on site at the same time as this project was implemented). The 'target group' includes local villagers residing near the project and those taking an interest in the implementation of the project.

5.6 Methods

This chapter assesses the work of a student–faculty trip to Abheypur, India in January 2010 including the implementation of a soak pit. To fully understand process interactions, researchers interviewed project participants and community members, including representatives of the University of Hartford student–faculty team, members of the Hartford professional chapter of EWB, and adult members of the local village. Interviews of fourteen participants in total (four members of the student–faculty team, three from the professional EWB chapter, and seven local villagers) were conducted using a semi-structured interview instrument. Questions investigate the core elements of the theory (motivation, knowledge and power) and were developed from a similar study focusing on the implementation of wetland restoration projects in Europe and the United States (Owens, 2008). Finding villagers for the study involved informal inclusion of people near the project site. This might represent a population more curious or invested in the project than a random sample from the village. That being said, the project is extremely localized, involving the development and construction of a soak pit that impacts the collection of water on one street: as a very local problem, it was apt to include people *using* the street to understand the project's impact. Slightly different questions were asked of the American and Indian counterparts, as necessitated by their roles in the process. The analysis and discussion of the professional engineer interviews is not included in this chapter, as they worked in Abheypur at a different time than the student–faculty team, and their primary project was not the soak pit.

Core variables

The independent variables in this project are the core actor motivation, knowledge and power. Chapter 3 uses different terminology (cognitions, resources) and also spends just a few paragraphs on this part of the theory before jumping into the larger contextual layers of influence (which are not used in this analysis). The theory works under the assumption that motivation, knowledge and power are not simply three important characteristics among others, but instead are critical variables in evaluating processes (Bressers, 2004; Owens, 2008).

Motivation is conceptualized as the way the implementation of the project plays a part in the achievement of an actor's objectives (Bressers, 2004). The actor's motivation incorporates whether or not the actor aligns with the implementer's goals, as well as sources of work-related motivation. It also includes external sources that can influence motivation, such as normative, economic, political and social influences (i.e. is there pressure from student–faculty team members, or other villagers to take part in, or not take part in, the project; is there an economic incentive to participate; is there pressure in either direction from politicians, or other societal influences). In addition, motivation incorporates one's attitude toward the implementation objective and one's attitude toward the other actors involved. Motivation also incorporates an actor's sense of self-effectiveness, or their confidence about their ability to implement (or stop the implementation of) the project.

Knowledge includes accessibility of materials and the transparency of the process for both targets and implementers. Knowledge-level measurement involves project awareness for relevant actors, understanding policy requirements and benefits, and knowledge of other stakeholders and their roles in the process. This measurement includes an assessment of the documentation available. It also incorporates the accessibility of knowledge about the project, as well as process complexity and uncertainties.

The conceptualization of power in this study incorporates three aspects: resources and informal and formal areas of control. Resources may involve financial, personnel or time support. These resources have the ability to strengthen or weaken the position of an individual, organization or agency. Formal power is that given to a group, individual or agency according to the law. Informal power stems from other channels (e.g. coalition building) and may balance or counteract formal power sources (Bressers, 2004). It is important to understand who has the formal power to realize the project, and to what degree, while also recognizing who actually exercises power in each case (Bressers, 2004). CIT analyses the balance of power between the implementer and target.

The dependent variable is the type of process interaction that is predicted to occur based on varying combinations of the independent variables, and can take one of many forms, including active cooperation, passive cooperation, forced cooperation, opposition, obstruction, joint learning or no interaction (see Bressers, 2004). In this way, using the CIT allows researchers to evaluate how core stakeholders interact in project implementation, as well as shed light on how successful (or unsuccessful) the given implementation is in reaching its collaborative goals. In this analysis, the format of the CIT is used to then apply Gajda's criteria of collaboration in order to better understand the collaborative nature of the process.

Data analysis

For motivation, the resulting score is the proportion of positive responses divided by the total number of responses on a scale of 0.0 to +1.0. When assessing interviews via the theory, motivation must be considered as positive, neutral or negative, which requires transforming the score to a scale of −1.0 to +1.0 and defining what delineates the three categories. To create scores of 'negative' and 'positive' distribution, 0.50 is subtracted from all original motivation scores, changing the scale of 0.0 to +1.0 to −0.50 to +0.50. Next, this −0.50 to +0.50 scale is multiplied by two and transformed to a scale of −1.0 to +1.0. Within this configuration, motivation can be thought of as:

−1.00 to −0.21 = negative motivation
−0.20 to +0.20 = neutral motivation
+0.21 to +1.00 = positive motivation

These intervals are unequal because of a desire to be conservative but also flexible in the estimation of the neutral category. This scale transforms interview results to fit the theory conceptualization. For knowledge, responses earn positive or negative points as they depict the level of knowledge held by each actor. The interview score is based on responses indicating positive levels of knowledge as a proportion of total number of relevant questions, on a scale of 0.0 to +1.0. The power score exists as a proportion of responses indicating sources of power within the process over the total number of relevant questions on power for each actor. As with the knowledge score, power exists on a scale of 0.0 to +1.0. Additionally, the two actor power scores are compared to each other to determine which actor holds the balance of power. Specifically, a difference of 0.0 to 0.14 between the two scores indicates that the power is balanced between the actors; a difference of 0.15 points or greater indicates that one actor holds the balance of power over the other actor. This estimates a difference in balance of power that is conservative but also flexible enough to allow for some small differences in power between actors. Both the positive/neutral/negative scale and the balance of power quantification were used and tested in an earlier study of forty-eight wetland restoration cases and have proven effective to date (Owens, 2008).

As the CIT is a two-actor model, actor-independent variable scores were merged, combining all of the members of the student–faculty team, and all of the villagers into one implementer and one target score, respectively. The theory was applied to determine the hypothesis predicted by the CIT and the dependent variable was measured by creating a case summary and an analysis of the roles of actors in the process. Finally, the case summary was compared to the CIT hypothesis to understand if the two were the same.

The goals of this chapter include evaluating this process for collaboration. This is accomplished by assessing the case in terms of Gajda and Koliba's (2007) traits of collaboration. Analysis compares the goals of collaboration with an expectation of how each trait might appear in this real-life context. Table 5.1 depicts the elaboration of these traits from Gajda and Koliba with a column that indicates how these traits may look in the context of this case.

Table 5.1 Gajda and Koliba's traits of collaboration and their application to this case

Trait	*Explanation*	*What would this look like in practice?*
Shared purpose	'Two or more entities come together for a reason – to achieve a vision or to do something – that could not be accomplished in isolation' (p. 29)	Implementer and target are not necessarily both positively motivated; one could be positive and one neutral. Or both positive
Cycle of inquiry: dialogue	'Collaboration implies some degree of person-to-person communication in any organizational setting, rendering dialogue an essential characteristic' (p. 31)	Evidence of person-to-person communication through meetings, discussions, opinion interviews and ongoing communication when university partners are not in India
Cycle of inquiry: decision making	In a high-functioning community of practice … all members actively determine the merit and worth of their practices and make choices about what to do and what to stop doing … Merely sharing ideas and swapping strategies is not enough to improve practice (p. 31)	Structure for democratic planning and decision making. Is there a mechanism for Indian partners to say 'We don't want this'?
Cycle of inquiry: action	Regardless of whether it is documented in writing or verbally expressed, a mutually agreed on decision or plan for action, by itself, leads to nothing. Community of practices, and the individuals within them, must take actions… (p. 31)	Action within the context of implementation by both university and Indian partners (i.e. volunteer hours from both groups)
Cycle of inquiry: evaluation	The extent to which a decision and action have merit or are worthwhile is determined through the systematic collection and analysis of performance data (p. 31)	Collecting data on projects now and into the future, from all relevant perspectives

5.7 Findings and discussion

Contextual Interaction Theory assessment

This analysis tests the following theory-derived hypothesis:

> If the student–faculty team is motivated to implement the soak pit, while the local villagers are positive or neutral about implementation, and the engineers have sufficient knowledge, then the interaction process will have the character of cooperation. When both actors are positive, active cooperation will occur.

Table 5.2 shows both the individual and compressed scores for implementers and targets in this process. Names of participants have been removed to protect privacy and maintain anonymity. Each actor is defined by their role in the process (i.e. implementers are members of the student–faculty team, and targets are local villagers). For motivation, the first column (Motivation) shows the number of pro-implementation responses based on the total number of questions asked. The second motivation column (M scale) includes these responses transformed onto the scale of –1.0 to 1.0 as described above. The third column (Category) indicates whether the overall score is positive, neutral or negative. The next column (Knowledge) shows the responses for each actor while the column (K scale) shows the transformation of those responses on a scale of 0.0 to 1.0. The column Power shows the actor's responses to questions about power, and the final column (P scale) indicates these responses on a scale of 0.0 to 1.0. The aggregate rows show the combined scores for all target (villagers) and implementer (student–faculty team) respondents.

In this case, the hypothesis predicts active cooperation. The unbalanced power never comes into play because both actors have motivation for the project. This prediction corresponds to both analysis and observations during implementation in India, where cooperation best describes the interaction. Interviews were performed at the end of the process; therefore this analysis captures one moment in time, and not any (potential) change over time. Using the CIT as an analytical tool, if the target group lacked motivation, the theory would hypothesize an interaction of forced cooperation. As researchers, it is important to understand if this is a case of cooperation/active cooperation as opposed to forced cooperation. In cases of knowledge transfer, forced cooperation, or coercion, may be typical. That being said, it is important to understand the nature of the cooperation in this case.

A qualitative assessment of the interviews allows some insight into whether the situation reflects forced or active cooperation. In reading the interviews of the student–faculty team, it is clear that the villagers were not initially cooperating. One Hartford faculty member states, 'We had this problem of you know, why isn't anybody volunteering to help?' Similarly, a

Table 5.2 Analysis results

Role	*Motivation*	*M scale*	*Category*	*Knowledge*	*K scale*	*Power*	*P scale*
Implementer:	(+) 5 of 5	1.00	positive	(+) 4 of 4	1.00	(+) 2 of 4	0.50
Student–	(+) 2 of 4	0.00	neutral	(+) 3 of 6	0.50	(+) 0 of 2	0.00
faculty team	(+) 2 of 2	1.00	positive	(+) 4 of 6	0.67	(+) 0 of 2	0.00
members	(+) 1 of 2	0.00	neutral	(+) 4 of 5	0.80	(+) 1 of 2	0.50
Implementer aggregate	**(+) 10 of 13**	**0.54**	**positive**	**(+) 15 of 21**	**0.71**	**(+) 3 of 10**	**0.30**
Target:	(+) 3 of 3	1.00	positive	(+) 2 of 2	1.00	(+) 0 of 3	0.00
Villagers	(+) 2 of 3	0.33	positive	(+) 1 of 1	1.00	(+) 0 of 2	0.00
	(+) 2 of 2	1.00	positive	(+) 2 of 2	1.00	(+) 0 of 2	0.00
	(+) 3 of 3	1.00	positive	(+) 2 of 2	1.00	(+) 0 of 1	0.00
	(+) 1 of 2	0.00	neutral	n/a	n/a	(+) 0 of 1	0.00
	(+) 1 of 2	0.00	neutral	(+) 1 of 1	1.00	(+) 0 of 2	0.00
	(+) 2 of 2	1.00	positive	n/a	n/a	n/a	n/a
Target aggregate	**(+) 14 of 17**	**0.65**	**positive**	**(+) 8 of 8**	**1.00**	**(+) 0 of 11**	**0.00**

student responds, 'At first they weren't... helping so we had to discuss it with them.' This may have partially had to do with the way the plan changed over the course of the trip. As one student says, 'If we got there ... and made everything clear and told them what we're doing ... that would have gotten them to help us from the beginning.' In other words, the student–faculty team noted that the local villagers were not assisting, and while they saw this as a problem, they did not attempt and could not coerce or force them into helping. Instead, it took some time for the villagers to 'pitch in' and help digging the soak pit. In many ways, these events demonstrate that cooperation was not forced, but instead that unwilling participants eventually became more willing. Why they became more willing is not completely clear. Unfortunately, any changes in motivation are not captured in this one-time analysis. Comparing the nature of the collaboration to an ideal provides a more holistic sense of the cooperation that took place.

In terms of problem-solving, the project is an example of an adequate transfer of knowledge from the developed to the developing world. While a soak pit may not provide the long-term benefits of more permanent infrastructure, it is a short-term solution that improves the daily lives of local villagers in Abheypur. In a village where the national and state governments are unable to meet demands for the improvement of roads and transportation, the faculty-student team project provides a stop-gap measure that improves local conditions. IWRM has been criticized for its ambitiousness, but for many developing countries, simple solutions provide much needed relief (see Chapter 2). While in total the goals of IWRM may seem unachievable, small projects throughout the world have merit and may over time meet the lofty goals of IWRM.

Collaboration assessment

Gajda and Koliba's hallmarks of collaboration associated with purpose, dialogue, decision making, action and evaluation (as presented in Table 5.1) are used to evaluate the project. Using Gajda's criteria to assess collaboration, there are indicators of purpose, dialogue and evaluation. In terms of purpose, both actor groups are positively motivated toward implementation of the soak pit. When considering dialogue, both meetings and interactions during the trip to India are taken into account, as well as the ongoing relationship of project planning and management with the Navjyoti India Foundation. There is also an adequate level of evaluation, primarily due to the EWB requiring follow-up, reports and self-evaluation within the context of their programme. Student–faculty teams return to India each year to implement new projects as well as evaluate and maintain past projects.

Areas that need improvement include decision making and action. In terms of decision making, there is not a formal structure in place for local villagers to say 'We don't want this.' Informally, of course, this is possible, but to engender stronger collaboration a more formal structure should be in place. This aspect of decision making is closely tied to the concept of action. Without volunteers and assistance from local villagers, these projects would fail to be implemented in the short-term project period. The qualitative assessment of this case shows that while action (volunteering) from local villagers was the ultimate result, it took several days to get off the ground. When critically examining the context of the transfer in terms of collaboration, this is a clear area for improvement.

To address these issues, the student–faculty team communicated with the Navjyoti India Foundation in an email dated 16 February 2010 (sent by project leader Marcia Hughes of the Center for Social Research, University of Hartford). The email provided goals for future work, including but not limited to the following recommendations:

- 'All projects need to be community-driven and based on a formalized decision-making process. For each effort moving forward, explicitly establish a shared purpose/role among [student–faculty team] members. Always consider the caste/subcaste/gender/poverty issues and related power/resources, knowledge, and motivation among different groups.'
- 'Make collaboration a goal (Navjyoti, Sarpunch/Panchayat, Block/District level, community level) ... We should ask the following questions: (a) Are all stakeholders involved? (b) Is there a shared purpose and understanding of the project? (c) Is there equal decision-making? Has the community been involved and/or surveyed for their feedback? (d) Is there local leadership for the project (i.e. in the village)? (e) Expect and plan for disagreement and conflict.'

- 'Streamline the EWB collaborative process into the following steps: (a) Develop project ideas with community leaders and Navjyoti. (b) Navjyoti can survey/scan community for input/feedback on project ideas. (c) Navjyoti can serve as link between EWB and village and district leadership (i.e. for communicating project design development, and planning). (d) Fine tune plan and develop budget-tasks and responsibilities (prior to starting project).'
- 'Still need monitoring/planning/modifications to make previous implemented projects stable and sustainable.'
- 'Over the next several years, monitor systems that have been implemented to ensure proper use and satisfaction with system(s), community leadership, and overall quality of water.'

Only future analysis can show whether these attempts to further encourage the collaborative nature of this partnership are successful.

5.8 Conclusions

When citizens of an industrialized country embark on water projects that transfer expertise and knowledge into the developing world, it is both a goal and an obligation to work in a collaborative way with local people. This analysis of the implementation of one water project focuses particularly on evaluating collaboration. Using the CIT allows insight into not only the outcome of implementation (cooperation), but also provide context for *how* and *why* actors are cooperating. It also allows analysts to make critical distinctions between the varied forms of cooperation (e.g. forced or active cooperation). This analysis reveals how the people involved in this project influence cooperation. It also reveals the weaknesses of this collaborative relationship, shedding light on ways to improve this work India.

This analysis shows that while the student–faculty team achieves some goals of collaboration, they are not yet fully collaborating according to Gajda's criteria. Issues of caste and gender might be improved by paying special attention to these issues in facets of collaboration such as dialogue and decision making.

Sustainability projects that transfer knowledge and expertise to the developing world are being promoted by developing and industrialized nations' governments, non-governmental organizations, universities and non-profit groups. This analysis asks whether one specific case of knowledge transfer, the implementation of a soak pit by a student–faculty team, meets criteria of collaboration. Though knowledge transfer may take on the tone of coercion, it is the responsibility of ethical and responsible researchers to at least recognize if coercion is taking place, and at most prevent it (Dolowitz and Marsh, 1996). Analysing implementation to understand how actors influence these interactions can inform these various organizations, researchers and evaluators about successes and pitfalls in the process. Using the CIT allows a systematic approach providing evidence to stakeholders about the implementation as well as the nature of collaboration. It can enable research and programme partners in the developed world to take responsibility for their role in projects, and to ensure that true collaboration takes place.

References

Ackermann, C. and Steinmann, W. (1982) 'Privatized policy-making: Administrative and consociational types of implementation in regional economic policy in Switzerland', *European Journal of Political Research*, vol. 10, no. 2, pp. 173–185.

Ball, B.P. (1976) 'Water pollution and compliance decision making', in C. Jones and R. Thomas (eds) *Public Policy Making in the Federal System*, Sage, Beverly Hills, CA.

Berman, P. (1980) 'Thinking about programmed and adaptive implementation: Matching strategies to situations', in H.M. Ingram and D.E. Mann (eds) *Why Policies Succeed or Fail*, Sage, Beverly Hills, CA.

Bressers, J.T.A. (2004) 'Implementing sustainable development: How to know what works, where, when and how', in W.M. Lafferty (ed.) *Governance for Sustainable Development: The challenge of adapting form to function*, Edward Elgar, Cheltenham.

Browne, A. and Wildavsky, A. (1984) 'Implementation as mutual adaptation', in J.L. Pressman and A. Wildavsky (eds) *Implementation*, University of California Press, Berkeley, CA.

Bunker, D.R. (1972) 'Policy sciences perspectives on implementation processes', *Policy Sciences*, vol. 3, no. 1, pp. 71–80.

Davies, T. and Mason, C. (1982) 'Gazing up at the bottoms: Problems of minimal response in the implementation of manpower policy', *European Journal of Political Research*, vol. 10, no. 2, pp. 145–157.

Dolowitz, D.P. and Marsh, D. (1996) 'Who learns what from whom: A review of the policy transfer literature', *Political Studies*, vol. 44, no. 2, pp. 343–357.

Edwards, G.C., III (1980) *Implementing Public Policy*, CQ Press, Washington, DC.

Engineers without Borders (2012) 'Our story', online, available at: www.ewb-usa.org/about-ewb-usa/our-story, accessed 17 September, 2012.

Fetterman, D.M. (2001) 'The transformation of evaluation into a collaboration: A vision of evaluation in the 21st century', *American Journal of Evaluation*, vol. 22, no. 3, pp. 381–385.

Fetterman, D.M. (2002) 'Empowerment evaluation: Building communities of practice and a culture of learning', *American Journal of Community Psychology*, vol. 30, no. 1, pp. 89–102.

Gajda, R. (2004) 'Utilizing collaboration theory to evaluate strategic alliances', *American Journal of Evaluation*, vol. 25, no. 1, pp. 65–77.

Gajda, R. and Koliba, C. (2007) 'Evaluating the imperative of intraorganizational collaboration: A school improvement perspective', *American Journal of Evaluation*, vol. 28, no. 1, pp. 26–44.

Gross, N., Giacquinta, J. and Bernstein, M. (1971) *Implementing Organizational Innovations: A sociological analysis of planned educational change*, Basic Books, New York.

Hill, M. and Hupe, P. (2002) *Implementing Public Policy*, Sage, London.

Larson, J.S. (1980) *Why Government Programs Fail: Improving policy implementation*, Praeger, New York.

Levitt, R. (1980) *Implementing Public Policy*, Croom Helm, London.

Nakamura, R.T. and Smallwood, F. (1980) *The Politics of Policy Implementation*, St. Martin's Press, New York.

O'Toole, L.J. (2000) 'Research on policy implementation: Assessment and prospects', *Journal of Public Administration Research and Theory*, vol. 10, no. 2, pp. 263–288.

Owens, K.A. (2008) 'Understanding how actors influence policy implementation: A comparative study of wetland restorations in New Jersey, Oregon, the Netherlands, and Finland', PhD thesis, Twente University Press, Enschede.

Pahl-Wostl, C., Craps, M., Dewulf, A., Mostert, E., Tabara, D. and Taillieu, T. (2007) 'Social learning and water resources management', *Ecology and Society*, vol. 12, no. 2, pp. 1–19.

Palumbo, D.J. and Calista, D.J. (1990) *Opening Up the Black Box: Implementation and the policy process*, Greenwood Press, New York.

Preskill, H. (2009) 'Reflections on the dilemmas of conducting environmental evaluations', *New Directions for Evaluation*, vol. 2009, no. 122, pp. 97–103.

Pressman, J.L. and Wildavsky, A.B. (1973) *Implementation: How great expectations are dashed in Oakland; Or, why it's amazing that federal programs work at all, this being a saga of the Economic Development Administration as told by two sympathetic observers who seek to build morals on a foundation of ruined hopes*, University of California Press, Berkeley, CA.

Raelin, J.A. (1980) 'A mandated basis of interorganizational relations: The legal-political network', *Human Relations*, vol. 33, no. 1, pp. 57–68.

Raelin, J.A. (1982) 'A policy output model of interorganizational relations', *Organization Studies*, vol. 3, no. 3, pp. 243–267.

Williams, W. (1980) *The Implementation Perspective*, University of California Press, Berkeley, CA.

Williams, W. (1982) 'The study of implementation: An overview', in W. Williams (ed.) *Studying Implementation: Methodological and administrative issues*, Chatham House Publishers, Chatham, NJ.

Wittrock, B. (1985) 'Beyond organizational design: Contextuality and the political theory of public policy', in K. Hanf and T.A.J. Toonen (eds) *Policy Implementation in Federal and Unitary Systems: Questions of analysis and design*, Martinus Nijhoff Publishers, Dordrecht.

6 Contextual considerations shaping the transferability of policies for drinking water source protection

A Canadian case study

Rob de Loë and Dan Murray

6.1 Introduction

From mega-cites to tiny hamlets, those responsible for providing drinking water in countries around the world are challenged to collect, treat and distribute safe water efficiently and effectively. Treatment systems are a critical component in most efforts to ensure safe drinking water. However, increasingly it is recognized that treatment systems alone cannot guarantee safe drinking water. Thus, the multi-barrier approach to drinking water safety has become an international norm (Davison *et al.*, 2005). In the multi-barrier approach, attention focuses not only on the treatment system, but also on the safety of the networks of pipes and pumps that distribute treated water to consumers, the effectiveness of monitoring regimes that ensure water safety at all points, and the quality of the water sources that supply the system (Islam *et al.*, 2011). Protection of source water quality is considered to be the first line of defence in the multi-barrier approach (Hrudey *et al.*, 2003).

In the context of drinking water supply, source water protection (SWP) conventionally involves a diverse range of activities designed to ensure that surface water and groundwater resources are protected from contamination. Land-use plans, policies and regulations created by municipalities can be used to control the location of activities that threaten water quality. Contaminant source inventories, monitoring programmes, water-related data collection, education programmes and spills contingency plans are other common tools used in SWP programmes (de Loë *et al.*, 2005).

The tools, techniques and procedures for SWP are well understood and have been documented for decades in a host of manuals and guidebooks (e.g. Jaffe and DiNovo, 1987; Davison *et al.*, 2005). Less clear are the most appropriate ways to organize the various actors to address the challenges of SWP. Globally, responsibility for implementation of SWP measures is divided among a wide range of actors including individual land owners, private firms, municipal governments, watershed management districts, state/provincial governments and national governments. However, the balance of responsibilities among actors varies from place to place. This diversity reflects the many differing political, cultural, socio-economic and institutional contexts that exist around the world; selected examples of different approaches to this challenge are presented in Box 6.1. The different approaches taken by these countries highlight the kinds of governance challenges that are associated with protection of source waters. What should the roles of state agencies be relative to municipalities? What is the appropriate balance among regulatory approaches versus non-regulatory approaches such as stewardship and economic incentives? Potential answers to these fundamental governance questions are highly context dependent.

Box 6.1 Selected examples of approaches to source water protection around the world

Source water protection in the *United States* is rooted in the 1996 amendments to the federal *Safe Drinking Water Act* (SDWA). Under this statute, each US state was required to develop and have approved by the federal Environmental Protection Agency (EPA) a plan as to how it would assess all source waters within the state. Once assessments were completed, appropriate strategies for protecting source waters were to be developed, with local governments often playing key implementing roles (de Loë and Lukovich, 2004).

In *England*, source water protection occurs in connection with numerous different legislative and policy instruments, coordinated by an array of government and non-government organizations, operating at a variety of different political levels (Watson *et al.*, 2009). Although no laws, policies, or programmes are tied explicitly to source water protection *per se*, several contribute to that goal. Examples include *The Action Programme for Nitrate Vulnerable Zones*, the European Commission's *Groundwater Directive*, pollution control regulations, and the *Catchment Sensitive Farming Programme*. The Water Framework Directive, with its requirements for member states to achieve 'good' ecological and chemical status by 2015, is of course a critical institution.

New Zealand's 2007 *Health (Drinking Water) Amendment Act* made mandatory the development and implementation of public health risk management plans for all drinking water supplies that service more than 500 people; plans are encouraged but not made mandatory for smaller water supplies (Ministry of Health, 2008). The Act is complemented by the National Policy Statement for Freshwater Management 2011 (Government of New Zealand, 2011) which specifies a key role for local government. These changes represent a considerable shift from the treatment-focused activities that dominated drinking water management in New Zealand in the past.

Jurisdictions that have not yet determined how to deal with a public policy challenge, or which are seeking to reform existing but ineffective policies, increasingly look to the experiences of other places for inspiration, and as a way to shorten the policy development cycle (Mamadouh *et al.*, 2002). However, the extent to which one jurisdiction's policies and other institutions can be used in another depends heavily on a series of contextual considerations relating to local political, social and economic circumstances. As is discussed in Chapter 2, this is well understood in the policy transfer literature (Dolowitz and Marsh, 2000; De Jong and Mamadouh, 2002), but seems to be less well understood among some government policymakers, who continue to seek de-contextualized 'best practices' (Stead, 2012). In the water domain, for example, Ingram (2011) notes that during the last five decades policymakers seeking standardized solutions to common problems have failed to take account of contextual consideration that limit the extent to which institutions can be transferred from place to place.

In this chapter, we explore the context within which a novel SWP system that has recently emerged in the Province of Ontario, Canada, is embedded. Ontario's system is relatively new compared to international examples such as those in Box 6.1. Nonetheless, despite occurring at the subnational level, Ontario's approach has the potential to become an internationally-relevant example that is considered for policy transfer because it attempts to address problems that are ubiquitous in the source water protection domain in a way that reflects new ways of governing grounded in collaboration. We argue that any jurisdiction seeking to adapt or adopt elements of Ontario's system must carefully evaluate the fit relative to several major contextual considerations. The contextual considerations that are critical to determining the transferability of this system are rarely captured in the kinds of international evaluations of 'best practices' that are common in the water sector (e.g. Australian Productivity Commission, 2000; Organisation for Economic Co-operation and Development, 2011).

In the balance of the chapter, we identify key contextual considerations that determine fit, provide a brief overview of Ontario's system for SWP, and evaluate the prospects for transferring elements of Ontario's SWP system to other jurisdictions. Unlike most of the other chapters in the book, our focus is not on where the ideas underlying Ontario's system originated (i.e. Ontario as a *recipient* of policy transfer). Instead, our analysis focuses on the kinds of considerations that people looking to Ontario as a *source* of new policies must take into account prior to engaging in policy transfer. In this respect our chapter provides the basis for what Mossberger and Wolman (2003) describe as 'prospective policy evaluation'.

6.2 Contextual considerations that shape policy transferability

Failed policy transfers are common, and reasons for failure are well understood in the literature (if not in practice). For example, Dolowitz and Marsh (2000) point to three common reasons for failed policy transfer:

1. *uninformed transfers* that occur when borrowing jurisdictions do not have enough information about policies and how they operate in their originating jurisdictions;
2. *incomplete transfers* that occur when crucial elements of a policy or programme are not implemented during transfer; and
3. *inappropriate transfers* that occur when critical differences between contextual considerations in the original and recipient jurisdictions are not appropriately considered.

Avoiding these problems demands a strategic approach grounded in awareness of and information about programmes and policies elsewhere, along with systematic assessment of those policies and programmes in their original setting; this provides a basis for evaluating their potential application in a new setting (Mossberger and Wolman, 2003).

In this chapter, our focus is on the contextual considerations that define the setting in which the policy currently exists. We characterize this as assessing the factors that will determine 'fit' (Swainson and de Loë, 2011). Understanding these factors permits evaluation of the fit of policies in new contexts. For analytical purposes, we organize the many factors that are known to influence the transferability of policies (and thus 'fit') under three broad domains: (1) the policy context; (2) the nature of the policy problem; and (3) the characteristics of the policy being transferred. These domains are briefly described below. Table 6.1 summarizes major implications for evaluation of fit.

Examination of the policy context includes consideration of the political and institutional, resources and capacity, cultural, and biophysical contexts of the original and receiving jurisdictions. All jurisdictions have distinct political systems comprised of bureaucratic structures and institutions, political ideology and existing policies and laws (Dolowitz and Marsh, 1996; Mossberger and Wolman, 2003). Similarly, financial resources, bureaucratic size and efficiency, and capacity for policy implementation are different in each jurisdiction (Dolowitz and Marsh, 1996), as are sociocultural norms and values (Mossberger and Wolman, 2003; Roland, 2004).

The biophysical context in which a policy or programme is situated also is important when considering the potential transferability of water and other environmental policies (Swainson, 2008; Michaels and de Loë, 2010). In the water field hydrological, ecological and biophysical conditions greatly influence the type of policies that can successfully be implemented (Malano *et al.*, 1999; Saleth and Dinar, 1999; Bandaragoda, 2000). The

Table 6.1 Indicators of policy transferability or 'fit' relevant to water governance

Factors	*Indicators of transferability ('fit')*
Policy problem	• Similar policy problems and policy goals in the source and receiving jurisdictions (Hospers and Beugelsdijk, 2002; Mossberger and Wolman, 2003)
Characteristics of the policy being transferred	• Policy is less complex with a single goal that results in limited changes (as opposed to multiple goals that result in significant or widespread change) (Rose, 1991; Hospers and Beugelsdijk, 2002) • Policy has fewer perceived externalities (as opposed to proposing with a wide-ranging impact) (Rose, 1991; Linos, 2006) • Policy is well theorized and has clearly defined the cause and effect structure (Rose, 1991; Hospers and Beugelsdijk, 2002; Linos, 2006)
Policy context	• The receiving jurisdiction has adequate financial resources, bureaucratic size and efficiency, and capacity for policy implementation (Dolowitz and Marsh, 1996; Evans, 2004) • Equivalent financial and other resources between source and receiving jurisdictions allocated for implementation (Hospers and Beugelsdijk, 2002) • Similar cultural institutions and values in the source and receiving jurisdictions (Hospers and Beugelsdijk, 2002; Linos, 2006) • Similar political and institutional contexts in the source and receiving jurisdictions (Dolowitz and Marsh, 1996; Mossberger and Wolman, 2003; Benson, 2009) • Similar biophysical contexts in the source and receiving jurisdictions (Malano *et al.*, 1999; Saleth and Dinar, 1999; Bandaragoda, 2000)

Source: Swainson and de Loë (2010).

likelihood of successful policy transfer is higher when the biophysical policy contexts in the original and receiving jurisdiction are compatible.

A key consideration when determining transferability is whether the nature of policy problem is similar in the original and receiving jurisdictions (Mossberger and Wolman, 2003). To determine transferability there must be consideration of the reasons why the policy was developed in the originating jurisdiction (i.e. the policy goals), and whether the opportunities and barriers to effecting change are similar in the original and receiving jurisdictions (Hospers and Beugelsdijk, 2002).

Finally, three characteristics of the policy being transferred influence transferability. First, policies that are less complex are more transferable than polices that are more complex (Rose, 1991; Dolowitz and Marsh, 1996). Second, and related to this point, policies with fewer perceived externalities are more likely to be transferable than those with a wide-ranging impact (Rose, 1991; Linos, 2006). Finally, the more clear the cause and effect relationships underlying a policy, the more transferable it will be (Rose, 1991; Dolowitz and Marsh, 1996).

Collectively, the factors presented in Table 6.1 provide a basis for 'screening' contextual considerations. Compatibility at the contextual level is necessary for successful policy transfer, we suggest, but never sufficient. As Ingram (2011) notes, the real world is far too complex to permit simplistic lesson-drawing. Therefore, systematic application of the factors in Table 6.1 simply permits what Mossberger and Wolman (2003) describe as 'prospective policy evaluation'. Put another way, consideration of these kinds of contextual considerations allows prospective transplanters to screen the suitability of models from other places (De Jong and Mamadouh, 2002).

6.3 Governance for source water protection in Ontario

The Province of Ontario (Figure 6.1) is Canada's largest province by population (with 13.2 million people in 2010) and second largest by land area (with just over one million km²). The majority of the population is concentrated in the southern portion of the province, with over 2.5 million

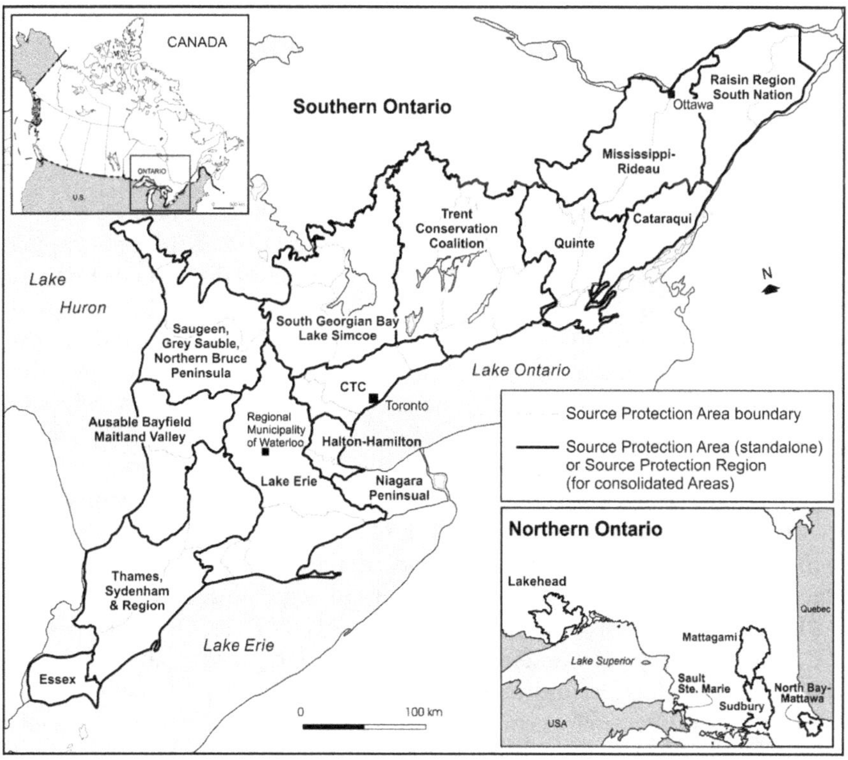

Figure 6.1 Southern Ontario source protection areas.

people in the City of Toronto alone (Government of Ontario, 2012). The vast majority of Ontario's residents (approximately 82 per cent) are served by drinking water treatment and distribution systems that draw on a mix of surface water and groundwater sources; the remainder, most of whom live in rural areas, are self-supplied, primarily by groundwater (Statistics Canada, 2011). Most drinking water systems in Ontario are operated by municipalities or public utilities, with some involvement of private firms, under a regulatory and policy framework created by the provincial government (Water Strategy Expert Panel, 2005).

In the decade leading up to 2000, the level of complacency regarding the safety of Ontario's drinking water system was generally high. This was most directly reflected in a weakening of the provincial government's regulatory environment and budget cuts to provincial agencies responsible for monitoring and regulating drinking water systems (Kreutzwiser and de Loë, 2010). Complacency changed to alarm in May 2000 when contamination of the water supply of the Town of Walkerton (population 4,800 people) in southern Ontario by *Escherichia coli* O157:H7 and *Campylobacter jejuniin* resulted in the deaths of seven people and over 2,300 serious cases of illness.

A public inquiry under the leadership of Justice Dennis O'Connor was launched in response to the incident. Justice O'Connor recommended that a multi-barrier approach to drinking water safety be adopted in Ontario, and that SWP– organized at the watershed scale and based on a collaborative approach – be the first line of defence (O'Connor, 2002b). In arriving at this recommendation, the Inquiry received oral and written submissions through public hearings and via comments submitted on its website; conducted its own research; commissioned twenty-five external research papers; and met with various experts. Justice O'Connor was also assisted by an expert advisory panel (O'Connor, 2002b). The text of the final report reveals the impact on O'Connor's recommendations of domestic (Ontario) and international experiences with watershed- or catchment-based management. Justice O'Connor's recommendation were accepted almost wholesale by the provincial government of the day, which started the process of creating a new SWP system with guidance from an Advisory Committee on Watershed-Based Source Protection Planning (2003). That government was defeated in the general election in 2003, but its successor remained committed and released its own white paper (Ontario Ministry of Environment, 2004), which provided the broad basis for the current legal framework for source water protection in Ontario, the *Clean Water Act* (Statutes of Ontario, 2006, Chapter 22) and its regulations.

The *Clean Water Act* defines the scope of SWP in Ontario. Where O'Connor (2002b) recommended a broader focus on the watershed as a whole, the Act focuses more narrowly on drinking water systems; self-supplied water users such as rural property owners served by their own wells are not covered. The provincial government (through the Ministry of the Environment) has ultimate responsibility for drinking water safety. However, new responsibilities for the identification of drinking water threats and for the development of policies to address these threats have been assigned to existing organizations. In addition, new institutions for collaboration (Source Protection Committees) based on watershed boundaries (Figure 6.1) have been created through the Act to aid source protection planning.

Ontario has a long history of watershed management through locally organized conservation authorities (CAs). The 1946 *Conservation Authorities Act* provided the means by which CAs could be created. Since 1946, thirty-six CAs have been established in Ontario. CAs are based on partnerships with municipalities, which provide the members of the CA boards. Under the *Clean Water Act*, the area over which each CA has jurisdiction is defined as a 'Source Protection Area'. The existing boards of CAs are then defined as 'Source Protection Authorities'. Recognizing the fact that CAs vary considerably in size and technical capacity, the *Clean Water Act* allowed for the consolidation of two or more Source Protection Areas into 'Source Protection Regions', with one of the member Source Protection Authorities designated as the lead. Nineteen of these regions are defined by regulations created under the Act: eight standalone Source Protection Areas, and eleven Source Protection Regions. To illustrate, in Figure 6.1 'Lake Erie' Source Protection Region is formed of four Source Protection Areas – with the largest one, the Grand River Source Protection Authority designated as the lead. In contrast, 'Niagara Region' is a standalone Source Protection Area.

In addition to these new institutional arrangements, the *Clean Water Act* created yet one more body: Source Protection Committees (SPCs). These are the basis for the broadly collaborative approach underlying source water protection in Ontario. For each of the nineteen regions in Figure 6.1, a SPC was created. Membership on SPCs is prescribed by Ontario Regulation 288/07, and includes a Chair along with representatives in equal proportions of municipalities; industrial, commercial and agricultural interests; and the general public. Where Indian Reserves defined under the federal government's *Indian Act* exist in the SPC's jurisdiction, provisions are made for seats for First Nations members.

Source Protection Committees are established by Source Protection Authorities, which then assist Source Protection Committees in performing their duties through providing scientific, technical and administrative support. Through a collaborative planning process, SPCs are required to create three major products that must be submitted to the Minister of Environment for approval:

1 a *Terms of Reference* to define the scope of the committee's activities;
2 an *Assessment Report*, which identifies watersheds and significant water features, characterizes water quality and quantity in terms of vulnerable areas for drinking water, and identifies activities that are or would be threats to drinking water; and

3 a *Source Protection Plan*, which provides policies for addressing drinking water threats identified in the Assessment Report and for monitoring those threats.

As of May, 2012, Assessment Reports were mostly completed and approved for the nineteen regions, and SPCs were working to produce detailed Source Protection Plans by the August 2012 submission deadline.

Following approval by the Minister of Environment, plans will be implemented by municipalities, provincial and other agencies, and land owners using both existing policy tools and new policy tools introduced by the *Clean Water Act* (such as the ability to prohibit certain activities in sensitive areas). For example, under the Act municipalities are required to develop and implement policies that reduce the risks to drinking water that are created by activities that occur within their jurisdictions; not only will they have to modify their own Official Plans to address significant threats identified in approved SPCs, but also they may require individual property owners to take actions that address these threats (*Clean Water Act*, Statutes of Ontario 2006, Chapter 22). Municipalities already have many of the tools they will need to address source protection needs (Plummer *et al.*, 2011). However, numerous new policies will be needed to address specific kinds of threats. For example, the Regional Municipality of Waterloo (RMOW) in southern Ontario (Figure 6.1), with a 2010 population of 492,390 people, is the largest community in Canada that is primarily dependent on groundwater (Statistics Canada, 2011). At the time of writing, the RMOW had just released its draft SWP policies for consultation. This document contains 318 policies that commit provincial ministries and agencies, regional and local municipalities to addressing existing and future significant threats to drinking water through policy tools that include regulation, prohibition, risk management planning, land-use planning, incentives and education (RMOW, 2012). Implementation, monitoring and coordination of these polices is likely to present significant challenges to the various stakeholders involved.

6.4 Prospects for transferring elements of Ontario's source water protection system

As suggested at the outset, Ontario is trying to solve a problem that is experienced around the world. Directly or indirectly, all drinking water systems are supplied from water bodies that are susceptible to contamination from adjacent and distant land uses and activities. These threats can be contemporary (e.g. agricultural or industrial activities that take place next to municipal drinking water wells or surface water intakes) or historical (e.g. long-forgotten buried wastes that leach into groundwater or surface water bodies) (Davison *et al.*, 2005). The roles that particular actors inside and outside of governments play in source protection vary widely from country to country. Nonetheless, SWP is an activity that inherently demands a multi-actor, multilevel approach to governance because no one actor is likely to have all the authority, resources or knowledge needed to address management tasks.

In light of the universal characteristics of the policy problem, and its growing significance around the world (Hrudey, 2011), it is not surprising that many jurisdictions are seeking to identify 'best practices' for drinking water safety (e.g. Australian Productivity Commission, 2000; United States Environmental Protection Agency, 2012). Several features of Ontario's approach may be attractive in jurisdictions seeking to establish a multi-actor, multilevel approach to governance for SWP. For example, Ontario's use of locally organized, collaborative bodies organized around watershed boundaries is consistent with global practices (e.g. Leach and Pelkey, 2001; Margerum and Whitall, 2004; Lubell, 2004; Bidwell and Ryan, 2006; Kallis *et al.*, 2009). From a more pragmatic perspective, Ontario's approach may be especially attractive to central governments that are seeking to share key responsibilities with local actors. For those interested in policy transfer *from* Ontario, what considerations will determine the potential 'fit' between Ontario's model and the receiving jurisdiction? Drawing on the factors identified in Table 6.1, this section highlights selected key considerations linked to the *policy problem*, the *characteristics of the policy being transferred* and the *policy context.*

Policy problem

The policy problem was defined almost entirely by the Walkerton outbreak. As noted previously, the commission of inquiry into the Walkerton contamination incident recommended a watershed-wide focus for SWP that took account of current and future threats to both water quality and water quantity (O'Connor, 2002b). However, the provincial government focused much more specifically on public health and safety. Thus, the *Clean Water Act* concentrates on the reduction of specific threats associated with wellhead protection areas and surface water intake protection zones for drinking water systems. This risk-based orientation has resulted in a highly data-driven, technical, regulated approach to the identification and management of threats to drinking water sources. For example, the work of SPCs is closely prescribed under Ontario Regulation 287/07 and in Director's Technical Rules that specify how Assessment Reports will be completed, what constitutes a threat to drinking water and how threats will be evaluated (Government of Ontario, 2009).

Importantly, the Act applies to all drinking water systems, including small ones that supply a limited number of households, but it does not cover private well owners. Their source protection needs, which are numerous, must be addressed through voluntary initiatives (Kreutzwiser *et al.*, 2011). Defining the policy problem in terms of communal systems was a pragmatic way to achieve economies of scale. An estimated three million people in Ontario are dependent on private wells (Kreutzwiser *et al.*, 2011), making each well owner, in effect, a drinking water system operator. In contrast, protecting the drinking water system of a municipality can increase water safety for thousands to millions of individuals.

Characteristics of the policy to be transferred

The system for SWP created in Ontario post-Walkerton is extremely complex. The overall goal is straightforward: reduction or elimination of threats to drinking water safety. However, SWP under the *Clean Water Act* creates new institutions for collaboration, increases demands for coordination and integration among existing organizations and activities, and adds to the responsibilities of existing organizations. At the same time, a host of new externalities are created, and considerable uncertainty exists regarding implementation and outcomes. Selected examples highlight this complexity.

Ontario's approach to environmental regulation historically has been quite prescriptive and top-down (Estrin and Swaigen, 1993). In some respects, the *Clean Water Act* continues the tradition of top-down regulation. For example, as noted above, the provincial government has strictly defined the mandates and roles of the SPCs, and what must be addressed in Source Protection Plans. At the same time, though, the Act gives responsibilities for plan development to locally organized SPCs, and makes a diverse range of actors outside the provincial government (e.g. municipalities, businesses, farmers) responsible for implementation of measures that reduce risks to drinking water safety.

SPCs are supposed to include representatives of all of the sectors that will be affected by their decisions. Nonetheless, there is no way to guarantee that all people in the represented sectors are in agreement with the decisions made by their representatives. For example, in the Lake Erie Source Protection Region (Figure 6.1), fifty-two municipalities located wholly or partly in the four Source Protection Areas that together comprise the Region are represented by seven municipal representatives on the SPC (Grand River Source Protection Authority, 2007). Thus, whether or not a key provincial government policy objective (increasing drinking water safety) is accomplished will be determined largely by the extent to which local actors accept the decisions made by the provincially mandated collaborative committees that represent them.

The degree of uncertainty created by the new approach to SWP in Ontario is very high. The policy is well theorized, in the sense that from a scientific and technical perspective, threats to source waters are well known and the kinds of actions that can be used to address those threats are well understood. For example, advice to groundwater-dependent municipalities seeking to protect the sources of their water supplies has been available for decades (e.g. Jaffe and DiNovo, 1987; Witten *et al.*, 1995). The real challenge, however, has always related to implementation. Following approval of the plans in 2012/2013, municipalities will be required to address significant threats to drinking water safety in wellheads and intake zones. How much resistance they will face, and what the impacts of these changes on land use planning and economic development at the local scale will be, are uncertain. The technical capacity of municipalities to address identified threats to source water is highly variable (de Loë *et al.*, 2002), which creates additional uncertainty.

A further source of uncertainty relates to the full costs of SWP. These costs are currently unknown, and may be unknowable. They are *unknown* because the costs of addressing the thousands of significant threats identified in Source Protection Plans have not been estimated. Given that many of these threats deal with long-established land uses, there is little reason to think that the costs will not be substantial. At the same time, some of the costs are *unknowable* because it is highly unlikely that all threats and vulnerabilities have been identified. To illustrate, every abandoned well is a potential threat to groundwater quality (Simpson, 2004). The Ontario Ground Water Association (2006) estimates that there are approximately 1.5 million improperly sealed wells in Ontario. Many of these wells are abandoned, and thus by definition their location (let alone their existence) is not known.

Policy context

The policy context for SWP in Ontario is multifaceted, and is defined by a host of financial, institutional, political, cultural and biophysical considerations. Key examples are addressed here, beginning with the fact that the Province of Ontario is vast in size and is extremely diverse and complex ecologically, hydrologically and geomorphologically (Ontario, 1984; MacRitchie *et al.*, 1994). Most of the provincial population is concentrated in southern Ontario (Figure 6.1), in a landscape that is dominated by agriculture, industry, suburban development and cities (Bone, 2010). The main implication of this biophysical and socio-economic diversity is that the specific threats to source waters, and the most appropriate ways to address those threats, will be highly localized. This in large part accounts for the key role played by the locally organized SPCs.

The costs of source water protection to date are considerable and must be taken into account by those seeking to transfer Ontario's model. For example, between 2004 and 2008 the provincial government committed approximately $CAD 120 million to support groundwater studies, technical assessments and plan development (Government of Ontario, 2006). An additional $CAD 7 million were allocated in 2007/2008 to support stewardship activities in rural areas (Government of Ontario, 2011). As suggested in the previous section, these expenses may be dwarfed by the actual costs to municipalities, conservation authorities, businesses and landowners associated with implementing the policies defined in each plan.

One of the defining features of Ontario's SWP system is its organization on a watershed basis. Watershed boundaries provide the basis for defining the jurisdictions of the Source Protection Areas and Source Protection Regions. Numerous problems exist when watershed boundaries are used as the basis for governance (see Cohen and Davidson, 2011). However, unlike many other jurisdictions, Ontario has a long history of watershed management through the CAs. In fact, Ontario's CAs were active proponents of watershed-based SWP at Justice O'Connor's Inquiry, and were key members of the Advisory Committee on Watershed-Based Source Protection Planning (2003). Therefore, in considering the transferability of Ontario's system it is essential to account for the fact that pre-existing watershed-based organizations with a long and generally successful track record of local collaboration and watershed management already existed in Ontario when the SWP system was conceived (Mitchell and Shrubsole, 1992). This is an important consideration because even though their size and capacity varied widely (Ivey *et al.*, 2002), many CAs already had in place elements of the technical expertise needed to undertake key SWP activities such as the watershed characterizations that are the backbone of the Assessment Reports mandated by the *Clean Water Act.*

Another key consideration that defines the policy context is the role played by municipalities. These actors will be critical to the success of SWP in Ontario because they are primarily responsible for implementation of the policies established in the Source Protection Plans. The capacity of Ontario's municipalities to undertake these activities is highly variable because they range in size from tiny rural townships to cities such as Toronto (population 2.5 million people). However, municipalities in Ontario have been responsible for land use planning and drinking water for well over a century (Estrin and Swaigen, 1993). As noted previously, while new policies will be required to implement Source Protection Plans, Ontario's municipalities can draw on an already strong base of powers and tools (Ivey *et al.*, 2006; Plummer *et al.*, 2011). Ontario's system of 'upper-tier' and 'lower-tier' municipalities is also an important consideration. This relationship has been effective at boosting the capacity of smaller communities located within the jurisdiction of larger upper-tier Regional Municipalities (where these exist). For example, the Regional Municipality of Waterloo (Figure 6.1) has an urban core of three larger cities that are surrounded by four large, rural townships. The rural townships have benefited from the sophisticated approach to groundwater protection implemented by the RMOW long before the Walkerton incident (Ivey *et al.*, 2006). Of course, the fact that CAs are a partnership with municipalities is another relationship that builds the capacity of both smaller and larger municipalities (Ferreyra and Kreutzwiser, 2007).

Finally, in terms of the policy context, it is essential to emphasize the impact of the Walkerton tragedy on public awareness regarding drinking water safety. As Christensen and Parfitt (2001, p. 3) note, 'Not only was the outbreak fraught with early allegations of negligence and wrongdoing, [but also] the speed at which it made people ill and, in a few tragic cases, killed them, was frightening.' The impact of the outbreak on the commitment of the provincial government of the day, and its successors, was equally dramatic. Prior to Walkerton, the erosion of concern for water issues in the provincial government was steady and growing (Kreutzwiser and de Loë, 2010). After the incident, the government of the day commissioned an independent inquiry that produced a damning report (O'Connor, 2002a) and a clear roadmap for SWP (O'Connor, 2002b), most aspects of which the Government of Ontario has implemented. Both general public and provincial government concern for drinking water safety have remained high in Ontario more than a decade later. A key test will be the extent to which the commitment to SWP remains strong during implementation of the plans.

6.5 Conclusions and implications

In countries around the world, the challenges associated with protecting the sources of water that supply drinking water to human populations are significant and growing. Countries such as the United States have a lengthy track record (Box 6.1). In contrast, Canada is a relative latecomer to SWP as a systematic policy objective (Simms *et al.*, 2010). Currently, Ontario is leading the way in Canada – due almost entirely to the shock it received from the Walkerton outbreak. This chapter revealed several features of Ontario's system that make it a distinctive and important example for an international audience:

- Reflecting the growing trend to multi-stakeholder collaboration for environmental governance, Ontario has developed a system where local actors representing relevant sectors develop plans using consensus-based decision-making. Planning occurs on a watershed basis, another widely accepted benchmark of good practice internationally.
- Lines of responsibility and authority are clearly delineated in Ontario's legislated system: the provincial government defines the rules and approves the plan; Source Protection Committees create the plans, working closely with municipalities, which will be responsible for implementing the plans.
- Integration of SWP objectives with land use planning and local economic development decision-making – a common gap around the world – is facilitated in Ontario via requirements for municipalities and individuals to actually change their practices once significant threats to source waters have been identified in approved plans. This alone is a marked departure from the more typical disconnect that exists in many jurisdictions between watershed management, on the one hand, and land use planning, on the other.

Policy transfer represents an opportunity to address common challenges through benefiting from the experiences of others. Ontario's approach was developed with knowledge of international experiences – but it is grounded much more strongly in the province's own lengthy history with watershed management. Assuming that implementation is successful, Ontario's system may become an attractive model to other jurisdictions that are seeking to address their own SWP challenges. However, realizing the benefits of policy transfer from Ontario will require that practitioners and policymakers understand the factors that shape transferability, and develop the capacity to utilize lessons appropriately in their own contexts. Numerous critical contextual factors that permit an initial screening of contextual compatibility were identified in this chapter. Some are so important that they alone may be the deciding factor. For example, the system conceived in response to the Walkerton outbreak builds directly on an almost six-decade partnership between municipalities and watershed-based CAs. Jurisdictions that lack similar watershed-based organizations, or whose local governments have not previously played a role in water management, should be very careful about which features of Ontario's model they transfer.

In conclusion, this chapter provides a cautionary tale for policy transfer. Water governance, by nature, involves multiple stakeholders operating at multiple scales. In the context of governance for SWP, institutional diversity is particularly complex. Furthermore, given the scales at which they operate, SWP systems are especially embedded in particular political systems, biophysical contexts, collective experiences and social norms. The policy transfer literature indicates that policies developed in such contexts are difficult to transfer, and this analysis supports that assertion. Hence, those seeking policy transfer from Ontario's system for SWP would be well advised to concentrate at the level of ideas that can be adapted to local circumstances, rather than considering wholesale transfer.

Acknowledgements

We gratefully acknowledge the support of the Social Sciences and Humanities Research Council, through funding for the project *Improving Water Governance Through Policy Transfer and Lesson Learning* (Rob de Loë, Principal Investigator). Additionally, we thank Charley Worte and Hugh Simpson for their comments on an earlier draft; Alyssa Roth for her background work; and Becky Swainson, whose contributions to a technical report (Swainson and de Loë, 2010) informed this chapter. Of course all errors or omissions are the responsibility of the authors.

References

Advisory Committee on Watershed-Based Source Protection Planning (2003) *Protecting Ontario's Drinking Water: Toward a Watershed-based Source Protection Planning Framework*, Ontario Ministry of the Environment.

Australian Productivity Commission (2000) *Arrangements for Setting Drinking Water Standards: International Benchmarking*, AusInfo, Canberra.

Bandaragoda, D.J. (2000) *A Framework for Institutional Analysis for Water Resources Management in a River Basin Context*, Working Paper 5, International Water Management Institute, Colombo, Sri Lanka.

Benson, D. (2009) *Review Article: Constraints on Policy Transfer*, CSERGE, School of Environmental Sciences, University of East Anglia.

Bidwell, R.D. and Ryan, C.M. (2006) 'Collaborative partnership design: The implications of organizational affiliation for watershed partnerships', *Society and Natural Resources*, vol. 19, no.9, pp. 827–843.

Bone, R. (2010) 'Ontario', in *The Regional Geography of Canada*, 5th edition, Oxford University Press, Toronto, ON, pp. 226–279.

Christensen, R. and Parfitt, B. (2001) *Waterproof: Canada's Drinking Water Report Card*, Sierra Legal Defence Fund, Toronto, Ontario.

Cohen, A. and Davidson, S. (2011) 'The watershed approach: Challenges, antecedents, and the transition from technical tool to governance unit', *Water Alternatives*, vol. 4, no. 1, pp. 1–14.

Davison, A., Howard, G., Stevens, M., Callan, P., Fewtrell, L., Deere, D. and Bartram, J. (2005) *Water Safety Plans: Managing Drinking-Water Quality From Catchment to Consumer*, Water, Sanitation and Health Protection and the Human Environment, World Health Organization, Geneva.

De Jong, M. and Mamadouh, V. (2002) 'Two contrasting perspectives on institutional transplantation,' in M. De Jong, K. Lalenis and V. Mamadouh (eds) *Institutional Transplantation*, Kluwer Academic Publishers, Dordrecht, pp. 19–32.

de Loë, R.C. and Lukovich, D. (2004) 'Groundwater protection on Long Island, New York: a study in management capacity', *Journal of Environmental Planning and Management*, vol. 47, no. 4, pp. 517–539.

de Loë, R.C., Di Giantomasso, S. and Kreutzwiser, R.D. (2002) 'Local capacity for groundwater protection in Ontario', *Environmental Management*, vol. 29, no. 2, pp. 217–233.

de Loë, R.C., Kreutzwiser, R.D. and Neufeld, D. (2005) 'Local groundwater source protection in Ontario and the Provincial Water Protection Fund', *Canadian Water Resources Journal*, vol. 30, no. 2, pp. 129–144.

Dolowitz, D. and Marsh, D. (1996) 'Who learns what from whom: A review of the policy transfer literature', *Political Studies*, vol. 44, no. 2, pp. 343–357.

Dolowitz, D.P. and Marsh, D. (2000) 'Learning from abroad: The role of policy transfer in contemporary policy-making', *Governance: An International Journal of Policy and Administration*, vol. 13, no. 1, pp. 5–24.

Estrin, D. and Swaigen, J. (1993) *Environment on Trial: A Guide to Ontario Environmental Law and Policy*, Emond Montgomery Publications Limited, Toronto, ON.

Evans, M. (2004) *Policy Transfer in Global Perspective*, Ashgate Publishing, Aldershot.

Ferreyra, C. and Kreutzwiser, R. (2007) *Integrating Land and Water Stewardship and Drinking Water Source Protection: Challenges and Opportunities*, Conservation Ontario, Newmarket, ON.

Government of New Zealand (2011) *National Policy Statement for Freshwater Management 2011*, Government of New Zealand, Wellington.

Government of Ontario (2006) 'Legislature passes Clean Water Act', 18 October, online, available at: www.news.ontario.ca/archive/en/2006/10/18/Legislature-Passes-Clean-Water-Act.html, accessed 28 May 2012.

Government of Ontario (2009) *Tables of Drinking Water Threats: Clean Water Act, 2006*, Government of Ontario, Toronto, ON.

Government of Ontario (2011) 'Ontario leading North America: Clean, safe, reliable drinking water', 29 July, online, available at: www.news.ontario.ca/ene/en/2011/07/ontario-leading-north-america--clean-safe-reliable-drinking-water.html, accessed 28 May 2012.

Government of Ontario (2012) 'About Ontario', online, available at: www.ontario.ca/en/about_ontario/index.htm, accessed 18 May 2012.

Grand River Source Protection Authority (2007) 'Memorandum to Members of the Grand River Source Protection Authority re Groupings for Municipal Representation on the Source Protection Committee. Dated August 31, 2007', Report No. SPA-08-07-04, online, available at: www.sourcewater.ca/swp_committee/Municipal_final.pdf.

Hospers, G.J. and Beugelsdijk, S. (2002) 'Regional cluster policies: Learning by comparing?' *Kyklos*, vol. 55, no. 3, pp. 381–401.

Hrudey, S.E. (2011) *Safe Drinking Water Policy for Canada*, C.D. Howe Institute, Toronto, ON.

Hrudey, S.E., Payment, P., Huck, P.M., Gillham, R.W. and Hrudey, E.J. (2003) 'A fatal waterborne disease epidemic in Walkerton, Ontario: Comparison with other waterborne outbreaks in the developed world', *Water Science and Technology*, vol. 47, no. 3, pp. 7–14.

Ingram, H. (2011) 'Beyond universal remedies for good water governance: A political and contextual approach', in A. Garrido and H. Ingram (eds) *Water for Food in a Changing World*, Routledge, Abingdon.

Islam, N., Sadiq, R., Rodriguez, M.J. and Francisque, A. (2011) 'Reviewing source water protection strategies: A conceptual model for water quality assessment', *Environmental Reviews*, vol. 19, no. 1, pp. 68–105.

Ivey, J.L., de Loë, R.C. and Kreutzwiser, R.D. (2002) 'Groundwater management by watershed agencies: An evaluation of the capacity of Ontario's conservation authorities', *Journal of Environmental Management*, vol. 64, no. 3, pp. 311–331.

Ivey, J.L., de Loë, R.C. and Kreutzwiser, R.D. (2006) 'Planning for source water protection in Ontario', *Applied Geography*, vol. 26, no. 3/4, pp. 192–209.

Jaffe, M. and DiNovo, F. (1987) *Local Groundwater Protection*, American Planning Association, Chicago, IL.

Kallis, G., Kiparsky, M. and Norgaard, R. (2009) 'Collaborative governance and adaptive management: Lessons from California's CALFED Water Program', *Environmental Science and Policy*, vol. 12, no. 6, pp. 631–643.

Kreutzwiser, R. and de Loë, R. (2010) 'Water security: Current and emerging challenges', in B. Mitchell (ed.) *Resource and Environmental Management in Canada: Addressing Conflict and Uncertainty*, 4th edition, Oxford University Press, Toronto, ON, pp. 207–237.

Kreutzwiser, R., de Loë, R., Imgrund, K., Conboy, M.J., Simpson, H. and Plummer, R. (2011) 'Understanding stewardship behaviour: Factors facilitating and constraining private water well stewardship', *Journal of Environmental Management*, vol. 92, no. 4, pp. 1104–1114.

Leach, W.D. and Pelkey, N. (2001) 'Making watershed partnerships work: A review of the empirical literature', *Journal of Water Resources Planning and Management*, vol. 127, no. 6, pp. 378–385.

Linos, K. (2006) 'When do policy innovations spread? Lessons for advocates of lesson drawing', *Harvard Law Review*, vol. 119, pp. 1467–1487.

Lubell, M. (2004) 'Collaborative watershed management: A view from the grassroots', *The Policy Studies Journal*, vol. 32, no. 3, pp. 341–361.

MacRitchie, S., Pupp, C., Grove, G., Howard, K.W.F. and Lapcevic, P. (1994) *Groundwater in Ontario: Hydrogeology, Quality Concerns, and Management*, CS-94011, Environment Canada, Gatineau, QC.

Malano, H.M., Bryant, M.J. and Turral, H.N. (1999) 'Management of water resources: Can Australian experiences be transferred to Vietnam? *Water International*, vol. 24, no. 4, pp. 307–315.

Mamadouh, V., De Jong, M. and Lalenis, K. (2002) 'An introduction to institutional transplantation', in M. de Jong., K. Lalenis and V. Mamadouh (eds) *The Theory and Practice of Institutional Transplantation*, Kluwer Academic Publishers, Dordrecht/Boston, MA/London, pp. 1–16.

Margerum, R.D. and Whitall, D. (2004) 'The challenges and implications of collaborative management on a river basin scale', *Journal of Environmental Planning and Management*, vol. 47, no. 3, pp. 407–427.

Michaels, S. and de Loë, R. (2010) 'Importing notions of governance: Two examples from the history of Canadian water policy', *American Review of Canadian Studies*, vol. 40, no. 4, pp. 495–507.

Ministry of Health (2008) *Drinking-Water Standards for New Zealand 2005 (Revised 2008)*, Ministry of Health, Wellington.

Mitchell, B. and Shrubsole, D.A. (1992) *Ontario Conservation Authorities: Myth and Reality*, Department of Geography, University of Waterloo, Waterloo, ON.

Mossberger, K. and Wolman, H. (2003) 'Policy transfer as a form of prospective policy evaluation: Challenges and recommendations', *Public Administration Review*, vol. 64, no. 4, pp. 428–440.

O'Connor, D.R. (2002a) *Report of the Walkerton Inquiry: Part One, The Events of May 2000 and Related Issues*, Ontario Ministry of the Attorney General, Queen's Printer for Ontario, Toronto, ON.

O'Connor, D.R. (2002b) *Report of the Walkerton Inquiry: Part Two, A Strategy for Safe Drinking Water*, Ontario Ministry of the Attorney General, Queen's Printer for Ontario, Toronto, ON.

Ontario (1984) *Water Quantity Resources of Ontario*, Ontario Ministry of Natural Resources, Toronto, ON.

Ontario Ground Water Association (2006) OGWA Presentation to the Clean Water Act, Written Submission and Verbal Presentation to the Clean Water Act Hearings, 21 August 2006, online, available at: www.ogwa.ca/cmsja1/index.php?option=com_content&task=view&id=31&Itemid=94.

Ontario Ministry of Environment (2004) *White Paper on Watershed-Based Source Protection Planning*, Queen's Printer for Ontario (Integrated Environmental Planning Division, Strategic Policy Branch Ministry of the Environment), Toronto, ON.

Organisation for Economic Co-operation and Development (2011) *Water Governance in OECD Countries: A Multi-Level Approach*, OECD Publishing, Paris.

Plummer, R., de Grosbois, D., de Loë, R. and Velaniškis, J. (2011) 'Probing the integration of land use and watershed planning in a shifting governance regime', *Water Resources Research*, vol. 47, no. 9, W09502.

RMOW (Regional Municipality of Waterloo) (2012) *Source Protection Plan Pre-consultation Policy Package*, Regional Municipality of Waterloo, Waterloo, ON.

Roland, G. (2004) 'Understanding institutional change: Fast-moving and slow-moving institutions', *Studies in Comparative International Development*, vol. 38, no. 4, pp. 109–131.

Rose, R. (1991) 'What is lesson-drawing?' *Journal of Public Policy*, vol. 11, no. 1, pp. 3–30.

Saleth, R.M. and Dinar, A. (1999) *Evaluating Water Institutions and Water Sector Performance*, World Bank technical paper No. 447, The World Bank, Washington, DC.

Simms, G., Lightman, D. and de Loë, R. (2010) *Tools and Approaches for Source Water Protection in Canada*, Water Policy and Governance Group, University of Waterloo, Waterloo, ON.

Simpson, H. (2004) 'Promoting the management and protection of private water wells', *Journal of Toxicology and Environmental Health, Part A*, vol. 67, no. 20/22, pp. 1679–1704.

Statistics Canada (2011) *Canada Year Book 2011*, Catalogue no. 11-402-XPE, Statistics Canada, Ottawa, ON.

Stead, D. (2012) 'Best practices and policy transfer in spatial planning', *Planning Practice and Research*, vol. 27, no. 1, pp. 103–116.

Swainson, R. (2008) *Examining Institutional 'Fit': A Case Study of Environmental Water Allocation in New South Wales, Australia*, MA thesis, University of Guelph.

Swainson, R. and de Loë, R. (2010) *Exploring the Role of Policy Transfer in Water Governance: A Discussion Paper*, Water Policy and Governance Group, Waterloo, ON.

Swainson, R. and de Loë, R. (2011) 'The importance of context in relation to policy transfer: A case study of environmental water allocation in Australia', *Environmental Policy and Governance*, vol. 21, no. 1, pp. 58–69.

United States Environmental Protection Agency (2012) 'Source water protection case studies', online, available at: http://water.epa.gov/infrastructure/drinkingwater/sourcewater/protection/casestudies/index.cfm, accessed 18 May 2012.

Water Strategy Expert Panel (2005) *Watertight: The Case for Change in Ontario's Water and Wastewater Sector*, Queen's Printer for Ontario, Toronto, ON.

Watson, N., Deeming, H. and Treffny, R. (2009) 'Beyond bureaucracy? Assessing institutional change in the governance of water in England', *Water Alternatives*, vol. 2, no. 3, pp. 448–460.

Witten, J., Horsley, S., Jeer, S. and Flanagan, E.K. (1995) *A Guide to Wellhead Protection*, American Planning Association, Chicago, IL.

Part II

Transferring an international concept

7 Translating water policy innovations into Kazakhstan

The importance of context

Farhad Mukhtarov

7.1 Introduction

Central Asia has been one of the most popular destinations for testing water policy ideas and instruments that have been developed and practised in the West (Rahaman and Varis, 2008; Wegerich, 2008, 2010). With the Aral Sea crisis and the crucial need for transboundary water allocation between the Central Asian states, the emphasis on good governance of water has been repeatedly emphasized (O'Hara, 2004; Rahaman and Varis, 2008; see also Chapter 2). One of the defining characteristics of water policy reform in Central Asia is the externally driven agenda for policy change dominated by donors and international non-governmental organizations (NGOs) (Rahaman and Varis, 2008). In this chapter, I investigate the process of water policy reform in the biggest country of the region, Kazakhstan. The institutional reform of water management in Kazakhstan has been based on three policy innovations advocated by international organizations, such as the World Bank (WB), the United Nations Development Programme (UNDP), and the United Nations Economic Commission for Europe (UNECE). These innovations include (1) the introduction of an integrated water resources management and water efficiency plan referred hereafter as the IWRM plan; (2) river basin councils (RBC); and (3) a piece of modern water legislation in the form of the Water Code. However, the translation of these policy innovations into the context of Kazakhstan has proven to be a bumpy ride with unplanned consequences and great complexities. I argue that the context of Kazakhstan, the history of its water and land management, and the characteristics of water policy and politics, such as the presence or absence of a forum for policy deliberation, are among the main drivers that have pushed these policy innovations in a certain direction. Overall, I argue that the context plays a key role in how Western-born innovations popular within the circles of international experts touch ground in developing and transition countries.

A few conceptual clarifications are in order. A policy innovation is defined here as a 'program, idea or practice that is new to the government adopting it' (Walker, 1969, p. 881). By context, I mean the historically pre-determined institutions, the power structures of water administration in place, as well as the politics of water management and governance. This fits the term of 'structural context' introduced in Chapter 3. Policy concepts supplied from an international policy arena to individual countries are contested, negotiated and appropriated by national and subnational political actors (Molle, 2008; Mollinga *et al.*, 2006). Thus, it is of crucial importance to study the context and the political process of such appropriation. This chapter does so for the case of Kazakhstan.

Various disciplines study the spread of policy innovations across countries. Conventional approaches hinge on concepts such as policy transfer, policy diffusion, lesson drawing, institutional isomorphism and policy learning. However, few studies explore the process and outcomes of the introduction of policy innovations to new contexts with due attention to the importance of contextual factors. In addition, the literature rarely discusses the importance of policy deliberation and the opportunity for various stakeholders to interact continuously at multiple levels of governance in order to come to a shared understanding of the problem at hand and articulate their preferences. I define policy deliberation as the existence of forums, mechanisms and infrastructure for a wide array of stakeholders to be able to engage in a meaningful debate on policy problems and possible solutions, which would have a bearing on the outcome of decision-making process. As Page (1996, p. 1) notes, 'public deliberation is essential to democracy, in order to ensure that the public's policy preferences – upon which democratic decisions are based – are informed, enlightened, and authentic'. The role of policy deliberation in public policymaking has rarely been discussed in connection to policy transfer.

An exploratory case study method was used for this research (Yin, 2003). This means that instead of testing a presupposed research hypothesis formulated in a cause–effect manner, the study explores the process of introducing policy innovations into the context of Kazakhstan with the goal of identifying important factors that influenced this process and its outcomes. Data was collected in August 2008 during an internship at the UNDP-IWRM project office in Almaty, Kazakhstan. A follow-up data collection period was pursued in 2009. Both primary and secondary sources of data have been used. A total of seven semi-structured interviews were conducted with key policymakers and experts in Almaty and Astana, including four face-to-face and three telephone interviews. The interviews lasted up to two hours and were digitally recorded or noted. The data from the interviews was supplemented with the analysis of policy documents, written opinions of important policy actors, and the materials collected from the UNDP-IWRM project office.

The outline of the chapter is as follows: Section 7.2 briefly introduces the theoretical emphasis of this chapter. Section 7.3 introduces the water sector of Kazakhstan and grounds the discussion of the need for policy reform. Section 7.4 discusses the process and pathways whereby the three policy innovations have been translated from the international discourse into Kazakhstan. Section 7.5 discusses the policy process with reference to the drivers of policy change and policy context in Kazakhstan and the final section draws conclusions and implications from this study.

7.2 Theoretical emphasis: discursive contestation and context

The process of cross-jurisdictional travel of policy ideas is very common nowadays. As discussed in Chapter 2 of this book, the literature gives various labels to the process whereby innovations, policies and ideas cross spatial or organizational borders: 'imitation, emulation and innovation' (Westney, 1987); 'institutional transplantation' (De Jong *et al.*, 2002); 'lesson drawing' (Robertson, 1991); 'institutional transfer' (Jacoby, 2001), or 'policy transfer' (Dolowitz and Marsh, 1996; Evans and Davies, 1999; Stone, 2004). One of the key approaches for studying the process of cross-jurisdictional travel of policy ideas is through the lens of 'policy transfer'. For the purposes of this book, policy transfer is defined in Chapter 1 as '*the interactive process by which actors use knowledge (about water management and/or governance) that was developed in one context (country or governance level) for the development of water management and/or water governance in another context*'. This definition stresses the importance of the type of boundary that knowledge crosses in the process of policy transfer, be it spatial, organizational or cultural. Moreover, this definition underlines the interactivity of the process of policy transfer and brings to the foreground the issues of non-linearity, contingency and politics of policy.

The focus on the non-linearity of the political process is important. This is the central tenet of the emerging approaches to water governance, such as *policy translation* (Freeman, 2009; Mukhtarov, 2012, 2008) and *contextual water governance* (Brugnach and Ingram, 2012; Ingram, 2011; Ingram and Lejano, 2010). The theoretical basis for such an argument originates in the postmodern philosophy of Michel Foucault (Flyvbjerg, 2001), and has been subsequently built upon by the students of policy as discourse (Hajer, 1995), policy as argument (Fischer and Forester, 1993), and more recently, policy as practice (Cook and Wagenaar, 2012). All of these approaches have emphasized the importance of context, contingency of policy processes and forcefully argued against prescriptive social science. Instead, social science is there to illuminate the complexity and trigger reflexivity, pay attention to the multiple ways of knowing the same subject, and to inherent uncertainties in policy processes and hence fallibility of any attempts at 'policy engineering'. Such studies draw attention to the fallacy of universal maxims and approaches, such as uncritical beliefs in panaceas (Ingram, 2011). Thus, the task of a water policy scientist shifts from the traditional one of discerning generic patterns and predicting and controlling policy phenomena to understanding the context and exploring the richness of the processes involved. Such an approach favours 'thick' case study methods, which is the major method used in this volume. In exploring the process of introducing policy innovations to Kazakhstan, I analyse the discursive contestation around the meanings of policy innovations concerned and the importance of context in shaping the outcomes of the policy reform.

7.3 Water policy in Kazakhstan: policy actors and reform

It is widely accepted that the water resources of Kazakhstan are poorly managed (Zimina, 2003; Aliakhasov *et al.*, 2007; Hannan, 2006). The most notable are the problems of industrial pollution of rivers and lakes; the shrinking of the Lake Balkhash, the second biggest lake in the region after the Aral Sea; the competition for water between hydroelectricity production and irrigation; very inefficient water use and large water losses in the pipelines and irrigation channels (UNECE, 2008). Little attention has been paid to water quality, and instead the emphasis falls rather on water quantity. Poor financial and human capacity is a serious constraint (UNECE, 2008). The main cause of the crisis is in the poor water management system and such problems as centralized administration, poor and over-bureaucratized communication among government agencies, and weak cross-sectoral coordination amidst common fragmentation of responsibilities (Zimina, 2003). Table 7.1 shows key institutions involved in water management in Kazakhstan.

Table 7.1 Key institutions involved in water resources management in Kazakhstan

Institutions	*Responsibilities*
Ministry of Environmental Protection	Issues environmental permits and monitors surface water quality and quantity through the national hydro meteorological institute. As of 2008, the territorial bodies of the ministry have been rearranged in accordance with the river basin principle.
Ministry of Agriculture	Hosts the major administration body for water resources – CWR. Responsibilities of the ministry, among others, include agricultural research, land reclamation, soil quality monitoring, drainage and salinity.
Provincial level administration	Appointed by the President, *akims* are heads of provincial territorial administration. They approve economic development plans, including those with an impact on water resources.
Municipalities	Municipalities are not elected but appointed, which creates a vacuum in the accountability of municipalities to the people they serve.
Interstate Commission on Water Coordination	Conducts water allocation between all five Central Asian states every year. CWR represents Kazakhstan in this organization, as well as in all intergovernmental talks.

Source: adapted from UNECE (2008).

According to Articles 37–40 of the Water Code, the Committee for Water Resources (CWR) placed within the Ministry of Agriculture is the main state agency charged with water-use planning and authorization. However, as UNECE (2008, p. 149) states:

> In practice, this task goes far beyond their (CWR's) capacity, at least with the current low staff levels and weak organization. This situation creates overlapping responsibilities and tensions between institutions, in effect the opposite of the better coordination sought through IWRM.

Placed within the Ministry of Agriculture, CWR is able to effectively oversee only the issues of irrigation and drainage. Being subordinate to the ministry, CWR is politically handicapped to push for any policy changes against the interest of its patron organization in this field (ADB, 2005). CWR lacks the political freedom to enact its formal 'strategic overview' role and, most importantly, the necessary capacity and resources (ADB, 2005; Aliakhasov *et al.*, 2007; UNDP-IWRM, 2007; Genina, 2007). In response to this situation, a number of international organizations proposed policy reform in the direction of IWRM as a possible solution (ADB, 2005; Aliakhasov *et al.*, 2007; UNDP-IWRM, 2007). The reform started with the introduction of new legislation. Discussed below is the process of introduction of these policy innovations into the context of Kazakhstan.

7.4 The introduction of the Water Code, IWRM plan and RBCs into the context of Kazakhstan

The introduction of the Water Code

The adoption of the Water Code has been an important step in the water policy reform in Kazakhstan. It has legalized the introduction of IWRM as well as river basin organizations to promote inclusiveness and public participation. The new Code also includes such concepts as water pollution fees and taxes, public participation and water user associations (Genina, 2007). However, critics have argued that little has been done to implement the innovations embedded in the new Code (Genina, 2007). The poor implementation of the Water Code indicates the insufficiency of the legal reform and the deeply entrenched character of informal institutions that determine policy. Alternatively, this may suggest the context insensitivity of those who prepared the Water Code.

The work on the Water Code started in 2002 and it was submitted to and approved by the parliament in 2003, though the process of issuing by-laws and making corrections has been ongoing (Mukhtarov, 2009). The preceding years have marked the changing water policy of Kazakhstan, namely, the passage of 'The Concept on Water Sector Development and Water Policy until 2010' (2002), and the 'Drinking Water 2002–2010' programme (2002), as well as the law on 'Water User Associations' passed in 2003 (ADB, 2005; Wegerich, 2008). Thus, the drafting of the Water Code came as part of this reformation wave. According to Genina (2007), the main instigators of the Water Code were the World Bank and the international experts who drafted the new law in cooperation with experts from CWR and a research institute under the Ministry of Agriculture. The Code has been developed with a strong influence from international actors and discourse. Three major pathways can be distinguished as directing such influence.

1 The final stage of the Water Code was partly funded from the budget of the Nura-Ishym River Management project sponsored by the United Kingdom's Department for International Development (UK DfID) with facilitation from the project manager, one of the key policy entrepreneurs in the Kazakh water policy reform (UNDP official, personal communication, 14 August 2008). This allowed for stronger articulation of the novel policy ideas in the document.
2 The Code incorporated experiences and debates from other countries and experiences from England, France, Spain, Germany, Russia and Poland were reportedly used by the working group (Kudaibergenuli, 2006).
3 The biggest inspiration for preparation of the Water Code was the Russian water code, as some sections of the two codes closely correspond to each other (V. Nee, personal communication, 2 February 2009; UNDP official, personal communication, 14 August 2008; independent expert, personal communication, 14 August 2008).

The importance of the context of Kazakhstan and the externally driven agenda for the preparation of the Water Code manifested itself in three ways. First, due to the limited time and funding from the UK DfID, the experts have been under pressure to produce the draft Water Code within a short time, and therefore, deliberately left it 'broad-brush' with many details to be specified by subsequent legal documentation. This has left the opportunity to adjust the by-laws to the needs to Kazakhstan, but the longer deliberation of the Water Code has become impossible given the time pressure. This is typical of water legislation in countries that embark on this with funding and technical assistance from donor countries and has also been noted in Afghanistan (Wegerich, 2010).

Another weakness of the Water Code is the unrealistic mandate it has given to CWR to coordinate water use by the various stakeholders indicated in Table 7.1. CWR possesses neither the resources, such as financial and human capital, nor the mechanisms, such as political and financial independence, from the Ministry of Agriculture for fulfilling this mandate. To be effective, the Water Code needs to be supplemented with administrative changes that would suit CWR with political, administrative and financial independence from the Ministry of Agriculture. Making such changes is a politically committing step and is indicative of the general trend of countries such as Kazakhstan to be willing to reform the legislation but reluctant to enact it. Finally, the Water Code introduced many new technical terms from the West, such as 'condominium' or 'water user association' without their linguistic and conceptual translation into Russian or Kazakh, or the preparatory work on explaining and institutionalizing this new policy vocabulary into the circles of water experts and practitioners. A more context-specific approach to drafting the law, but also to preparing the water community for these changes needs to be taken if the new water legislation is to produce actual changes.

The introduction of the national IWRM plan

The importance of the context into which IWRM has been introduced in Kazakhstan can be illustrated with the discursive richness that this term has taken in this country. IWRM is viewed by policy actors in Kazakhstan in three different ways: (1) as a *process of management* that is new to Kazakhstan and needs to be established from scratch through a comprehensive legal and institutional reform; (2) as a *managerial addition* to the old system of 'river basin plans'; and finally (3) as completely *identical to the 'river basin plans'*. Such diversity of meanings suggests the complex norm-formation and discursive politics that took place in the setting. This is also a manifestation of the process whereby a pre-given meaning of a policy innovation, the IWRM plan in this case, did not retain itself in a new policy context but overlapped with other meanings that already existed in the setting.

IWRM was introduced to Kazakhstan as part of the UNDP Project 'Preparation of the National IWRM and Water Efficiency Plan for Kazakhstan'. The project was initiated jointly by the Norwegian and the Kazakh governments, supported by the UNDP, GWP and UK DfID, and therefore was a product of international cooperation from the beginning (UNDP, 2008). It has resulted in the preparation and eventual approval of the plan by the State Budget Programme of Kazakhstan for 2009–2011 (A. Nikolaenko, personal communication, 16 August 2008; V. Nee, personal communication, 2 February 2009). The plan recommended restructuring the relevant governmental agencies (the CWR's increased role), introduction of RBCs, the national information system for monitoring of water use and quality, the principle of cost-recovery, improvements in water efficiency, capacity building and education programmes. The preparation of the plan has been influenced by knowledge from outside Kazakhstan through three pathways:

1 international experts within the UNDP-IWRM project, UK DfID and external consultants;

2 the GWP guidelines, as it was formally part of the project through the Central Asian GWP office;
3 since the plan developed upon the Water Code, which introduced the idea of RBCs, indirect influence came from the Russian water code and the French river basin organizations.

(V. Nee, personal communication, 2 February 2009; UNDP official, personal communication, 14 August 2008; Kazgiprovodkhoz official, personal communication, 15 August 2008.)

However, in spite of this significant external influence, the context of Kazakhstan has shaped how IWRM eventually came to be understood. Opinions varied as to what IWRM represented. For example, Dukhovny and Sokolov (2003), the representatives of the Central Asian branch of the GWP, see IWRM as a system for water management. Their view is fully reflected in their comments on the draft IWRM plan consultation document (UNDP-IWRM, 2007). Others believe that IWRM is equal to the already introduced river basin management plans. In the 1970s, the Soviet Union prepared river basin management plans that are still being used in Kazakhstan and Russia, albeit on a somewhat smaller scale. Called the 'complex schemes of use and protection of water resources',[1] those basin level plans include the inventory of all water and related land objects and socio-economic trends. In addition, since 1986 there have been eight Basin Water Authorities (BWAs) in the country, signifying that river basin management had already before this time been introduced in Kazakhstan. This led the government of Kazakhstan and some independent experts to claim that Kazakhstan has been complying with IWRM for the last twenty years (Kazgiprovodkhoz official, personal communication, 15 August 2008). Indeed, the process and the content of preparation of the schemes is pretty much the same as in Soviet times, except for the chapter on IWRM (Kazgiprovodkhoz official, personal communication, 15 August 2008). The previous manager of the UNDP-IWRM project, Alexander Nikolaenko, has defined the 'new version of IWRM' as adding the environmental and participatory elements to the previously practised 'schemes'. Thus, the diversity of opinions on what IWRM represents and how to implement it coexist in Kazakhstan, where there is limited opportunity for the multiplicity of stakeholders to engage in a debate to bring their viewpoints into the open. In order to partially address this issue, a new form of participatory water governance has been introduced in Kazakhstan, namely RBCs.

The introduction of river basin councils

The introduction of RBCs in Kazakhstan is regarded as the UNDP project's most successful element which is desirable for replication in Central Asia (Ryabtsev, 2007; Aliakhasov *et al.*, 2007; Hannan, 2006). This policy innovation required the BWAs to take the lead in the establishment of RBCs and arrange the preparation of basin agreements among its stakeholders. RBCs are advisory bodies and do not have statutory powers. According to Petrakov, a legal expert involved in the preparation of the Water Code (personal communication, 16 August 2008), the proposal to grant statutory powers to the RBCs had been rejected on the ground that a governance function cannot be granted to a civil society organization (which is the legal form of RBCs, according to the Civil Code of Kazakhstan). The membership of an RBC consists of the heads of a BWA (chairperson), *akimats* (provincial executive power) and *maslikhats* (provincial legislative assemblies). Local users, including representatives of other province-level executive agencies responsible for water management, NGOs and water user associations, and journalists are also invited to RBC meetings (Aliakhasov *et al.*, 2007).

In the process of preparing to set up RBCs, many obstacles had to be overcome (Hannan, 2006). The concept of public participation was totally new for Kazakhstan. In the beginning of the process, the national government (CWR and the Ministry of Agriculture), the NGOs and the BWAs all opposed the idea of RBCs (Hannan, 2006). The BWAs, who would initially carry the heaviest workload to get the RBCs going, protested most actively, while the NGOs were sceptical about 'another project without a real impact on the ground' (UNDP official, personal communication, August 14 2008). The positive side of RBCs was the creation of a forum for NGOs and ministries to discuss important issues. However, much support is needed for the new forms of governance, such as the RBCs as is exemplified by the episode with the Balkhash-Alakol RBC presented in Box 7.1.

Box 7.1 River basin councils in Kazakhstan, the example of Balkhash-Alakol RBC

The example of the pilot RBC in the Balkhash-Alakol basin, located in the southeast of the country illustrates formidable problems in sustainable functioning of councils. The pilot council organized five meetings, starting in the summer of 2008. Most of them were well attended, and the council established itself as one of the most progressive among the eight introduced. However, the lack of funding made its work difficult. According to the representative of the Balkhash-Alakol RBC (RBC official, personal communication, 13 August 2008):

> The members of the RBC are interested to participate, we put some questions for discussion, and we try to get them to participate and make input. The members of our Council are the state bodies, the executive bodies, the representative bodies, and the large water users, as well as the NGOs who participate passively, whereas the others are active. The last two RBC meetings of the

> council we organized without support from the UNDP, and there was a good turnout at the meetings! We have a large territory, and the whole basin is about 1,000 km to drive. And the NGOs are not interested to pay for themselves... We see the RBC as a chance to solve the problem on the spot and avoid writing complaint letters to each other.

The effectiveness of RBCs boils down to the capacity of Basin Water Authorities (BWAs), the interest and incentives of the RBC members, and the possible policy implications of that work. According to T. Hannan (personal communication, 14 October 2008), it will require at least five years for RBCs to start functioning on a sustainable basis, and ten years for the first results on the ground to be visible.

7.5 Discussion: the role of context and policy deliberation in Kazakhstan's water policy

Without making discursive diversity explicit through deliberative processes, the normative consensus and practical implementation of policy innovations will face predictable problems of misunderstanding, miscommunication and false expectations. Such deliberation can be introduced in the form of consultation processes in the preparation of policy documents, policy discussions at conferences and public hearings as well as issue-specific discussion in more focused forums, such as RBCs. The importance of policy deliberation has been manifested by the IWRM plan. As mentioned earlier, prior to the commencement of the UNDP-IWRM project, which formally introduced IWRM planning, there were some fifteen international IWRM-related projects in Kazakhstan. Those projects have introduced a new language to the Kazakh water policy community. Although this has not translated into a uniform understanding of the concept, familiarity with the term 'IWRM' has increased since then. However, there is no single dominant interpretation of IWRM and, therefore, no normative notion of what is good and therefore should be implemented. As Tirtishniy (2005) states, the Global Water Partnership promoted IWRM in Kazakhstan, but it did not suggest *how* to make it work (Tirtishniy, 2005). There is an unresolved tension between the modern notion of IWRM as equal to the schemes of river basin management in the 1980s and the *postmodern* notion of IWRM that embodies public participation, equable consideration of environment and economic interests (ICWE, 1992). This is an important testimony to the discursive richness of water policy. Such diversity needs to be made explicit through the creation of various venues where policy deliberation can take place.

The lack of deliberation can also be observed in the case of the introduction of the Water Code. The new Water Code has marked the discursive change as new terms and concepts have now been legalized. The new governance terms in Kazakhstan's Water Code include 'integration', 'decentralization', 'condominium' and 'water user association'. Not only are the new terms introduced as part of the discursive change, but also new terms are created by merging Western and Russian scientific terms. The new terms introduced without care form a divide within the policy community and cannot serve the common purpose if they are not addressed in detail and explained in special forums (Genina, 2007). This may create the feeling of inferiority among experienced experts, because they are not able to understand the new Water Code, as reported by interviewees. Such confusion with terms is a sure way towards stalled dialogue and may result in undemocratic decision-making. Once again, the absence of the forum for public discussions and consultations has inhibited the process of meaningful introduction of the Western policy vocabulary in Kazakhstan.

Finally, the case of RBCs illustrates the challenges of introducing such deliberation in the context of Kazakhstan. The legitimacy of this instrument lies within the assumption of river basins being a natural unit for water management and the necessity to include all the relevant stakeholders in decision making. However, as Box 7.1 illustrates, the globally accepted institution of the RBC has not been sustainably institutionalized in Kazakhstan. The government of Kazakhstan needs to initiate the more deliberative process of policymaking as the donor-funded projects are inevitably limited in funding and time period of operation and may not have actual policy impact. The short-term funding, the great territories that RBC members need to travel to attend meetings and the strong executive power of *akims*, representatives of the presidential power in cities and districts, have all undermined the basic idea of the RBC to be inclusive and promote democratic decision-making. The global concept of the RBC met the local context in Kazakhstan, which resulted in RBCs struggling for their survival in terms of funding, independence and sustainability of operation. Interestingly, the CWR controls the councils through its regional organizations, BWAs. Thus, the task of organizing participatory water management is in the hands of the old engineering elite that has practised the hydraulic mission (Allan, 2003) for decades. Such a situation relates to the structural context of Kazakhstan and its importance in shaping RBCs. In the civil service and organizational culture of Kazakhstan, much remains from the administrative-command system of the Soviet Union, which implies a top-down style of management in which policy deliberation is virtually absent and its introduction is met with resistance. In the context of Kazakhstan, I cannot agree more with Wegerich (2010) who states the following in a study of water legislation reform in Kazakhstan: 'priority should be on enabling local communities. Stringent legal frameworks might create more mistrust towards an already weak government, which in any case would not be able to enforce them in the medium term.'

7.6 Conclusions

The process of introduction of three policy innovations to Kazakhstan, namely, the new Water Code, the national IWRM plan and RBCs is presented in this chapter. I discussed the importance of the context in the ensuing policy processes. As a result of this study, it was established that IWRM and river basin management are being institutionalized in Kazakhstan. However, the process has been caught in contestation, negotiations and competing interpretations of these policy innovations. This is especially important as the UNDP considers new projects to spread the experience of Kazakhstan to Kyrgyzstan and Tajikistan, creating therefore a model for further policy translation in the region (Hannan, personal communication, 14 October 2008).

Three key findings emerged from this chapter. First, international discourses, actors and funding have been instrumental in putting the three policy innovations on the policy agenda of Kazakhstan. Second, these external drivers of policy change alone do not explain the modification that the policy innovations have experienced in Kazakhstan, and the manner in which national and subnational policy actors engaged with these ideas. The policy context, the pre-existent discourses and institutions in Kazakhstan and the interests of policy actors have been as important as the external drivers of the policy reform. Third, the importance of deliberation of policy innovations in the context of Kazakhstan cannot be overstressed, but the challenges of introducing such deliberation in a country context of Kazakhstan are daunting. The absence of a forum for open and inclusive discussions of the Water Code, the IWRM plan and RBCs inhibited public deliberation of these policy innovations before their enactment. Public deliberation has been further inhibited by a large number of alien policy terms with Western origins that found a way into the Water Code. The IWRM plan has not provided an opportunity to clarify the multiplicity of meanings attached to IWRM and the institution of RBCs proved dependent on external financial support, as the struggle for survival of the Balkhash-Alakol RBC has demonstrated. Development of capacity for policy deliberation, however, is a long-term process, and the inclusiveness of RBCs is an important step in this direction, especially considering the rarity of such initiatives in countries with a post-Soviet context.

Overall, reflexivity is necessary in the process of development and application of policy innovations in new policy contexts and attention to the politics of water governance in individual countries is important. While this is certainly not the first call for the greater attention to contextual issues in applying policy innovations developed and practised abroad, as well as not the first plea for more policy deliberation in applying them, there is still much to be done in that regard in Kazakhstan as well as other countries. These issues are especially important for countries where the water policy reform is externally imposed.

Acknowledgements

I would like to thank all interviewees and colleagues at the UNDP Integrated Water Resources Management project office in Almaty, Kazakhstan for assistance in this research. I am grateful to Helen Ingram, Kai Wegerich, Rahaman Mizanur and the editors of this book for valuable comments on earlier versions of this chapter.

Note

1 In Russian this corresponds to the 'Схема комплексного использования и охраны водных ресурсов'. These schemes have been practised since the 1970s, and according to several expert interviewees, these were the guiding documents for the Syrdarya and Amudarya river development that finally led to the Aral Sea disaster. However, according to technocrats, the schemes were technically impeccable and deviation from their implementation was what caused the disaster.

References

ADB (Asian Development Bank) (2005) 'Institutional strengthening of the committee for water resources', ADB, Almaty.

Aliakhasov, Z., Nikolaenko, A. and Petrakov, I. (2007) 'Water Resources Management in Kazakhstan: History, Current State, Analysis, Comparisons, Tables, Schemes, Recommendations', independent review, UNDP-IWRM, Almaty [in Russian].

Allan, T. (2003) 'IWRM/IWRAM: A new sanctioned discourse?' Occasional Paper 50, School of Oriental and African Studies, SOAS Water Issues Study Group, London.

Brugnach, M. and Ingram, H. (2012) 'Ambiguity: The challenge of knowing and deciding together', *Environmental Science and Policy*, vol. 15, no. 1, pp. 60–71.

Cook, N. and Wagenaar, H. (2012) 'Navigating the eternally unfolding present: Toward an epistemology of practice', *The American Review of Public Administration*, vol. 42, no. 1, pp. 3–38.

De Jong, M., Lalenis, K. and Mamadouh, V. (2002) *The Theory and Practice of Institutional Transplantation: Experiences with the Transfer of Policy Institutions*, Kluwer Academic Publishers, Dordrecht.

Dolowitz, D. and Marsh, D. (1996) 'Who learns what from whom: A review of the policy transfer literature', *Political Studies*, vol. 44, no. 2, pp. 343–357.

Dukhovny, V. and Sokolov, V. (2003) 'Lessons on co-operation building to manage water conflicts in the Aral sea basin', UNESCO IHP, Paris.

ICWE (International Conference on Water and the Environment) (1992) *The Dublin Statement on Water and Sustainable Development*, Dublin.

Evans, M. and Davies, J. (1999) 'Understanding policy transfer: A multi-level, multi-disciplinary perspective', *Public Administration Review*, vol. 77, no. 2, pp. 361–385.

Fischer, F. and Forester, J. (1993) *The Argumentative Turn in Policy Analysis and Planning*, UCL Press, London.

Flyvbjerg, B. (2001) *Making Social Science Matter*, Cambridge University Press, Cambridge.

Freeman, R. (2009) 'What is "translation"?', *Evidence and Policy*, vol. 5, no. 4, pp. 429–447.

Genina, M. (2007) 'The development of a new water code in the Republic of Kazakhstan', *Politikon*, vol. 13, no. 1, pp. 21–31.

Hajer, M. (1995) *The Politics of Environmental Discourse: Ecological Modernization and the Policy Process*, Oxford University Press, Oxford.

Hannan, T. (2006) 'The process of preparing a national IWRM and water efficiency plan for Kazakhstan', UNDP Water Governance Facility, Stockholm.

Ingram, H. (2011) 'Beyond universal remedies for good water governance: A political and contextual approach', in A. Garrido and H. Ingram (eds) *Water for Food in a Changing World*, Routledge, Abingdon.

Ingram, H. and Lejano, R. (2010) 'Transitions: transcending multiple ways of knowing water resources in the United States', in D. Huitema and S. Meijerink (eds) *Water Policy Entrepreneurs: A Research Companion to Water Transitions around the Globe*, Edward Elgar Publishing, Cheltenham.

Jacoby, W. (2001) 'The imitation-innovation trade-off: Does "borrowing dull the edge of husbandry"?' *Comparative Political Studies*, vol. 34, no. 3, pp. 263–293.

Kudaibergenuli, K. (2006) 'International experience for River Basin Councils', UNDP Project Documentation.

Molle, F. (2008) 'Nirvana concepts, narratives and policy models: Insights from the water sector', *Water Alternatives*, vol. 1, no. 1, pp. 23–40.

Mollinga, P., Dixit, A. and Athukorala, K. (2006) *Integrated Water Resources Management Global Theory, Emerging Practice, and Local Needs*, Sage, New Delhi.

Mukhtarov, F. (2008) 'Intellectual history and current status of integrated water resources management: A global perspective', in C. Pahl-Wostl, P. Kabat, J. Möltgen (eds) *Adaptive and Integrated Water Management: Coping with Complexity and Uncertainty*, Springer, Heidelberg.

Mukhtarov, F. (2009) 'The hegemony of integrated water resources management: A study of policy translation in England, Turkey and Kazakhstan', PhD thesis, Central European University at Budapest, Hungary.

Mukhtarov, F. (2012) 'Rethinking the travel of policy ideas: Policy translation in the water sector', *Policy and Politics*, online, available as a pre-press fast track article at: www.ingentaconnect.com/content/tpp/pap/pre-prints/15799;jsessionid=1frvpse9zxz69alice.

O'Hara, S. (2004) 'Central Asians divided over use of dwindling water supply', OSI: Local Governance Brief, Budapest.

Page, B. (1996) *Who Deliberates? Mass Media in Modern Democracy*, University of Chicago Press, London.

Rahaman, M. and Varis, O. (2008) 'Central Asian waters: Social, economic, environmental and governance puzzle', TKK-WD-03, Helsinki University of Technology, Helsinki.

Robertson, D. (1991) 'Political conflict and lesson-drawing', *Journal of Public Policy*, vol. 11, no. 1, pp. 55–78.

Ryabtsev, A. (2007) 'Report at the National Forum of River Basin Councils', Almaty.

Stone, D. (2004) 'Transfer agents and the global networks in the "transnationalization" of policy', *Journal of European Public Policy*, vol. 11, no. 3, pp. 545–566.

Tirtishniy, E. (2005) *Comments and Suggestions to the Concept Note for the National IWRM and Water Efficiency Plan*, Atasu, Almaty.

UNDP (United Nations Development Programme) (2008) 'National Integrated Water Resources Management (IWRM) Planning and Transboundary Dialogue in Central Asia', online, available at: http://europeandcis.undp.org/public/show, accessed 2 February 2012.

UNDP-IWRM (United Nations Development Programme Project on Integrated Water Resources Management) (2007) 'Draft Programme on Integrated Water Resources Management and Water Efficiency in Kazakhstan', UNDP-IWRM, Almaty.

UNECE (United Nations Economic Commission for Europe) (2008) 'Environmental Performance Review: Kazakhstan', UNECE, Geneva.

Walker, J.L. (1969) 'The diffusion of innovations among American states', *The American Political Science Review*, vol. 63, no. 3, pp. 880–899.

Wegerich, K. (2008) 'Blueprints for water user associations' accountability versus local reality: Evidence from South Kazakhstan', *Water International*, vol. 33, no. 1, pp. 43–54.

Wegerich, K. (2010) 'The Afghan water law: "A legal solution foreign to reality"?' *Water International*, vol. 35, no. 3, pp. 298–312.

Westney, E. (1987) *Imitation and Innovation: the Transfer of Western Organizational Patterns to Meiji Japan*, Harvard University Press, Cambridge, MA.

Yin, R. (2003) *Case Study Research: Design and Methods*, Sage, Thousand Oaks, CA.

Zimina, L. (2003) 'Developing water management in the South Kazakhstan', in A. O'Hara (ed.) *Drop by Drop: Water Management in the Caucasus and Central Asia*, OSI: Local Governance Initiative, Budapest.

8 Public participation as an essentially contested concept

Insights from water management in Turkey

Gül Özerol, Aysun Özen Tacer and Mine Islar

8.1 Introduction

Many countries undertake efforts towards participatory water management that consist of replacing top-down institutional structures with bottom-up and networked structures. These efforts can be explained by the instrumental and normative benefits expected from public participation, and the resulting promotion of 'public participation' by both international organizations and water scholars (see also Chapter 2 of this book). Debates around public participation indicate however that there are multiple views regarding the definitions and uses attributed to the concept. Gallie (1956) introduced the idea of an 'essentially contested concept' and identified the following five conditions that determine whether a concept is essentially contested:

- the concept denotes a valued achievement
- the achievement is internally complex although its worth is defined for the whole
- the achievement is ambiguous with different descriptions for different people
- the achievement is vague since its attribution can change contextually and temporally
- the users of the concept defend the way that they use it against the other users

In light of its multiple definitions and uses and general agreement with the above five conditions, public participation is considered to be an essentially contested concept. Several scholars have addressed the contested nature of public participation and the implications of this contestation. Day (1997) examines public participation in policy planning processes and identifies public participation as an essentially contested concept that reflects the nuances and difficulties of democracy and requires expertise even to grasp its complexity. Regarding public participation in water management, Sneddon and Fox (2007) touch upon the contestation of the concept. They examine a case in transboundary water management and observe that the multitude of national and international actors employ different meanings of public participation and use those meanings to pursue individual goals or to exert power over other actors. Taking a critical stance, Molle (2008) identifies public participation as a 'nirvana concept', meaning that, just like integrated water resources management (IWRM), public participation appears desirable and uncontroversial, whereas at the same time it conceals the politics of water management and can be exploited by certain groups for legitimizing their agendas. Molle further argues that the implementation of public participation relies on the essential roles of the state in empowering public members and of the experts in providing information, hence reflecting the power asymmetries instead of balancing them.

Thus, we can assert that public participation in water management often has a contested nature. However, the underlying mechanisms and the resulting implications of the contestation process in a particular time and space remain a research gap. These mechanisms and implications are particularly relevant in explaining the transfer of public participation to contexts with a history of non-participatory water management, since the contested nature of public participation can constitute an obstacle against effective transfer, i.e. the achievement of instrumental and/or normative benefits through introducing public participation. Accordingly, in this chapter we aim to improve the understanding about the contested nature of public participation by analysing the underlying mechanisms and the implications of contestation. In particular, we reflect on the following research questions:

- Why does the concept of public participation become contested in water management?
- What are the implications of the contested nature of public participation in water management on the effectiveness of its transfer to non-participatory settings?

The outline of the chapter is as follows. In Section 8.2, we describe how we conceptualize the contested nature of public participation. Then in Sections 8.3 and 8.4, we present our empirical findings from case studies in Turkey, where we examine the transfer of participatory approaches to the national framework of water management and to three cases at the local level. In Section 8.5, we discuss the similarities and differences of transfer to local cases. Finally, in Section 8.6, we revisit our research questions and draw on conclusions.

8.2 Conceptualizing the contested nature of public participation

Since the definitions and uses of a concept are used to indicate its contested nature, we choose to make use of a discourse analytic perspective in this chapter. Discourse analysis constitutes an appropriate approach as it gives the opportunity to understand what shapes a social reality through examining the *text* and the *context* (Phillips and Hardy, 2002). The *text* includes the written and verbal communicative messages, whereas the *context* refers to the circumstances within which the *text* is generated. Analysis of discourses can be particularly useful for revealing the mechanisms through which the phenomena in the environmental domain are socially constructed and contested (Hajer and Versteeg, 2005). Being a key social phenomenon of environmental decision-making, public participation can thus be effectively analysed from such a discourse analytic perspective. As Table 8.1 shows, we formulate three questions to analyse the public participation discourses in water management.

Answering the three questions in Table 8.1 reveals the discursive positions of actors regarding the meanings and uses that they attribute to public participation. The second column of the table includes discursive positions that the multiple actors can possibly hold depending on the actual empirical settings. The first question addresses the discourses about the scope of the participants, which relates to representativeness and equity of public participation. The scope of the participants usually consists of a subset of the public, since involving everyone is often infeasible or unrealistic under time, money and human constraints (Mostert, 2003; Creighton, 2005; Özerol and Newig, 2008). However, contestation can arise when some actors, who are advantaged in terms of power and access to resources, become increasingly mentioned as stakeholders. In such cases, disadvantaged

Table 8.1 Questions regarding discursive positions about public participation

Questions	*Possible discursive positions*
1 Who is 'the public'?	• All individual and collective actors are potential/actual participants. • Representation is biased towards the dominant/advantaged actors. • The scope of 'the public' is implicit and/or ambiguous.
2 What is 'the participation'?	• Participation is an element of decision-making processes. • Participation is symbolic ('business-as-usual' prevails). • The scope of 'the participation' is implicit and/or ambiguous.
3 Why was public participation introduced?	• Public participation is an instrument to improve policy outcomes. • Public participation is expected to contribute to the functioning of democracy. • Public participation is an obligation imposed by external actors. • The rationale behind public participation is implicit and/or ambiguous.

actors such as women, smallholders and the poor are often excluded from the scope of participants as they lack the necessary power and resources. The second question relates to the discourses regarding the level or type of participation. In her highly cited paper, Arnstein (1969) identifies several types of participatory mechanisms as 'non-participation' or 'tokenism'. These types of participation imply that either participation does not take place at all or it remains symbolic by not affecting the decision-making processes. In all types of participation, the scope of participation can become contested when actors attribute divergent or ambiguous definitions to it and use those definitions to further their interests. Therefore having clear boundaries for both the 'public' and 'participation' can improve the effectiveness of the transfer of public participation (Bressers and Lulofs, 2010). Finally, the third question addresses the discourses about the normative and instrumental rationales behind the transfer of participation (Innes and Booher, 2004). Unless the discourses about the rationale of participation contradict each other, multiple discursive positions can be present.

According to the Contextual Interaction Theory, contextual factors influence policy implementation processes by influencing actor characteristics (see Chapter 3 of this book). As the context is a key element of discourses, we apply this premise of the Contextual Interaction Theory to analyse actors' discourses regarding public participation. Accordingly, we assume that the discursive positions of different actors result from their characteristics (cognitions, resources and motivations) and are influenced by the wider, structural and case-specific contextual factors. Although they can vary with empirical cases, these positions address whether there is disagreement, uncertainty or ambiguity regarding the definition and use of 'public participation' in a given case. If there are such positions, it can be inferred that public participation is contested. As we mentioned above, the contested nature of public participation constitutes an obstacle against its effectiveness. We further argue that the more contested public participation is, the more difficult it is to effectively transfer it. This argument relies on the idea that a high level of disagreement, uncertainty or ambiguity in discursive positions leads to a lower likelihood of achieving the benefits expected from public participation. The implications of the contested nature of public participation on its effective transfer can be unfolded by further investigating the discursive positions of different actors.

8.3 Public participation and water management in Turkey

Turkey is a relevant case to analyse the transfer of public participation to water management since many attempts are being made towards a transition from non-participatory to participatory management. Prior to the 1980s, Turkish public management in general, and water management in particular, was characterized by a top-down and centralist institutional structure (Kibaroglu and Baskan, 2011). Since then, however, the country has been undergoing fundamental changes, which include among others the introduction of participatory mechanisms through the involvement of non-governmental actors in the planning, development and management of water resources. These changes can be attributed to various drivers such as the negotiation process to become a European Union (EU) member, the World Bank supported liberalization of the water sector, and the ongoing debates about water, energy and food security under the pressures of economic growth, population rise and climate change (Kibaroglu *et al.*, 2009).

Previous experiences with public participation in the broader sphere of environmental decision-making have been limited in Turkey, yet they are increasing. For instance, several recent official documents and legislation refer to public participation. The latest national development plan (SPO, 2006) pinpoints the importance of enhancing local decision-making capacity as well as knowledge sharing and partnerships between the public and private sector and non-governmental organizations (NGOs). The preparatory reports of the development plan, which are compiled by expert groups and NGO representatives, also mention public participation as a key element to incorporate into environmental decision-making (SPO, 2007a, 2007b). Along the same lines, the Environmental Law that was amended substantially in 2006 incorporates several environmental policy integration principles, including among others the participation of non-governmental actors in the decision-making processes. Another key legal document related to public participation in environmental decision-making is the Regulation on Environmental Impact Assessment (EIA), which was enacted in 1993 and revised three times as part of the EU harmonization process. Building on the EU EIA Directive, the regulation includes provisions regarding public consultations during the planning of large-scale infrastructure projects that can have significant environmental impact.

Despite the supportive outlook in policies and legislation, the fact that Turkey did not ratify the Aarhus Convention (UNECE, 1998) shows that access to information and public participation is lacking not only in water management, but also in all areas of environmental decision-making. Nevertheless, there are efforts to incorporate new participatory mechanisms and improve the implementation of the public participation processes. For instance, with support from the Dutch government and the European Commission, pilot projects have been implemented in western Turkey to establish the institutional background for the implementation of the EU Water Framework Directive, and public participation and knowledge dissemination have been identified as being among the most crucial topics for compliance with the requirements of the directive (Grontmij, 2005).

The agricultural sector has the highest share of national water use and agricultural water management has been substantially reformed since the

1990s (Kibaroglu *et al.*, 2009). With financial and technical assistance from the World Bank, the Directorate General of State Hydraulic Works (*Devlet Su İşleri*, hereafter DSI) has undertaken the so-called 'accelerated transfer programme', which was inspired by the Mexican model of irrigation management transfer (Molle, 2008). This programme led to the establishment of water user organizations (WUOs) that took over the responsibility of implementing 'participatory irrigation management'. Most of the WUOs in Turkey are irrigation associations or irrigation cooperatives, which are, in principle, responsible for surface water and groundwater, respectively (DSI, 2012). Although these two types of WUOs have legal and administrative differences, they are both structured around the idea of farmers' participation in irrigation management. This idea continues to be one of the key messages of DSI, which argues that farmers' awareness of irrigation and self-organization abilities enables them to internalize irrigation management, and thus ask for and contribute to infrastructure investments (DSI, 2011). The countrywide establishment of WUOs constitutes a substantial step towards increasing farmer participation and is considered a success by other countries and funding agencies (Vermillion, 1997; Merrey *et al.*, 2007). However, several issues such as the capture of WUOs' top management by the local elite and the inadequate monitoring of WUOs' performance impede effective implementation (Kadirbeyoğlu, 2008; Kibaroglu *et al.*, 2009; Özerol, 2013).

Similarly to the developments related to agricultural water management, hydropower development in Turkey has been under pressure to increase the involvement of non-governmental actors, particularly the private companies that came into play over the last decade (Kibaroglu *et al.*, 2009). The current government has embarked on an effort to promote small-scale hydroelectric facilities. The number is massive, with over 1,600 of such facilities being planned for Turkey's rivers over the coming decades. With the legal reform in 2001, use rights to the rivers are granted for forty-nine years to private companies to build small-scale, run-of-the-river hydroelectric facilities. Such a transfer of rights from the state to the private sector led to considerable tensions over rivers. The participation of lay public members is one of the contested issues since many communities were not satisfied with the outcomes of participatory meetings during the EIA of hydropower projects. According to Scheumann *et al.* (2011), the public consultations that are part of the EIA procedure only receive interest due to the significance of the social impacts on the local communities and the resettlements that result from the hydropower projects.

The experiences of the last three decades show that the transitions taking place in agricultural water management and hydropower development are characterized by a top-down imposition of participatory mechanisms led by the successive governments and supportive international agencies. The provisions in the relevant planning and legislative documents adopt an inclusive and participatory discourse. In practice, however, the NGOs can only minimally contribute and the lay public members are totally excluded from decision-making processes (Kibaroglu *et al.*, 2009).

8.4 Case studies of public participation in Turkey's water management

Turkey has seven subnational regions that differ in terms of geographical and climate conditions as well as cultural and ethnic backgrounds. To better understand how factors at the case-specific level influence public participation, we examine the following three cases:

1. South-Eastern Anatolian Region: the case of participatory irrigation management in the Harran Plain, where abundant surface water is transferred from the Euphrates River for irrigated agriculture.
2. Central Anatolian Region: the case of participatory irrigation management in the Konya Closed Basin, where scarce groundwater is extracted from wells for irrigated agriculture.
3. Black Sea Region: the case of public consultation in EIA during small-scale hydropower development in the eastern part of the region.

We selected these regional cases for their political and social relevance for water management in Turkey. These regions also fall into the geographic focus of our ongoing research regarding water management in Turkey. To collect the discursive data about public participation in these cases, we used three types of data sources, namely documents, interviews and direct observation. The documents include previously generated text in the form of legislation, official reports, and scientific reports and articles. Through interviews and direct observation, we generated data that did not exist in the documents, which are produced mainly by corporate actors such as central and local administration units, professional organizations and NGOs. The opinions of certain actors, such as the water users and the middle-level personnel of governmental organizations, are not directly reflected in most of the documents. Therefore, we conducted semi-structured interviews to elicit the perspectives of such actors about the public participation processes and we made direct observations during our visits to the case study sites. For each case, we first examined the discourses to reveal the definitions and uses that different actors attribute to public participation. Then we investigated whether, how and why these attributions were contested through examining the disagreements, uncertainties or ambiguities in the discursive positions of the actors involved. In the following subsections, we present the description of the cases, followed by the discourses that were uncovered regarding public participation.

Harran Plain

Harran Plain is located in Sanliurfa, one of the nine provinces in the South-Eastern Anatolian Project (*Güneydoğu Anadolu Projesi*, hereafter GAP). GAP is a multi-sector regional development programme that has been operating since the 1980s. Extensive irrigation is practiced in the plain using the Euphrates River's waters, which are stored in the Atatürk Dam reservoir and transferred to the fields through the 26.2 km-long Sanliurfa Tunnels. A large-scale canal system irrigates an area of around 1,500 km^2 (Aküzüm *et al.*, 1997). When the irrigation with water from the Euphrates started in 1995, the DSI stipulated the establishment of irrigation associations in the newly irrigated areas. Irrigation associations implement participatory irrigation management by operating and maintaining the irrigation canals, distributing water to the farmers and collecting the irrigation fee, which is intended to cover the expenses incurred by the WUOs. The DSI supports the irrigation associations through providing trainings, lending machinery and equipment, and giving technical advice. As the irrigated areas are expanded, new irrigation associations are established in the GAP region.

Who is 'the public'?

In 2005, irrigation associations acquired the legal status of *local administration associations* under the Ministry of Interior through the Local Administration Associations Law. This provision implied the involvement of local administrators, namely the heads of villages and the mayors of municipalities, as the natural members of the top management of the irrigation associations. The Irrigation Associations Law, which was specifically crafted for the irrigation associations and enacted in March 2011, redefines the legal status of the irrigation associations as *public legal entities.* The enactment of a specific law and the change in the legal status of the irrigation associations imply a retreat of local administrations away from the management of irrigation associations, and a re-regulation of the management structure of the irrigation associations through specialized legal provisions.

The Irrigation Associations Law also contains a provision about assigning the number of votes according to the size of the land owned or used by the members of the irrigation association and grants a maximum of five votes to the farmers that own or use larger areas of land. Such a provision favours the participation of large-scale farmers over the disadvantaged farmers, whereas there are no mechanisms to facilitate the participation of disadvantaged farmers. Harran Plain is one of the regions that will be affected the most from this provision since the land is unequally distributed and the landlords already have additional power in the management of irrigation associations. A smallholder farmer summarizes his views on being involved in management: 'You need to be *aga* [landlord] to manage things'.

Although it is among the mandates of participatory irrigation management, it is difficult to argue that the establishment of irrigation associations in Harran Plain enabled the involvement of disadvantaged groups in the decision-making processes. On the contrary, the hierarchical tribe structure continues to shape the management of irrigation associations (Özerol, 2013).

What is 'the participation'?

Both the DSI and the irrigation associations focus on the participation of *individual* farmers, which mainly occurs through their payment of the irrigation fee each year and voting in the irrigation association elections every four years. When it comes to the social interactions of farmers both with the top management and the staff of irrigation associations, the persistence of tribal culture, which relies on strong ties among the members of large families, is substantial (Özerol, 2013). The informal relationships are exploited in order to get things done by any type of organization: 'If you want to be served by the government offices, you need uncles [relatives or friends to ask for help]'. The tribal culture also leads to the prioritization of kinship and friendship relations over official relationships. This creates an uncertainty as to how management processes work and obstructs the enforcement of sanctions on the individual farmers and the irrigation associations when they violate the rules of irrigation management.

Regarding the enforcement of sanctions, the positions of farmers and DSI are relevant. Among the farmers, a common opinion about the functioning of irrigation associations is that 'irrigation associations just distribute the water'. Many farmers also think that the irrigation associations waste the fees that they collect from the farmers. However, until the enactment of the Irrigation Associations Law, the irrigation associations were not accountable to farmers, since the farmers were not members of the associations. The legal status of irrigation associations has also been an issue for DSI, since it lacked the legal power to enforce sanctions against the irrigation associations that neglect the maintenance of irrigation and drainage canals. Being local administration associations, the irrigation associations were administratively and financially accountable to the Ministry of Interior, which lacked the technical capacity to monitor the administrative and financial performance of the irrigation associations. With the enactment of the Irrigation Associations Law, the individual farmers gained the right to become members of irrigation associations, and DSI became the main administrative and technical auditor of the associations. Thus, the law changes the legal background for the accountability of irrigation associations to farmers and DSI. However, the enforcement of provisions towards improving the monitoring of administrative and technical performance remains to be seen.

Why was participation introduced?

The DSI has transferred the operation and management responsibility of almost all the irrigation schemes it constructed (DSI, 2012). In the preamble of the Irrigation Associations Law, the key rationale behind the transfer programme is expressed as

> the need to decrease the public expenditure on the operation, maintenance and repair of irrigation infrastructure, to improve the quality and speed of services, and to increase the collection rate of irrigation fees through creating the sense of ownership of farmers regarding the irrigation facilities

Such expectations are valid in many transfer programmes implemented throughout the world (Vermillion, 1997). In the Turkish case, all actors highlight the realization of economic expectations. However, there are criticisms about the lack of ownership by the farmers, since the irrigation infrastructure is deteriorated in many areas, including the relatively new infrastructure in the Harran Plain. A related item on the agenda of all related governmental organizations is to train the farmers in better irrigation management practices. If training were crucial to improve farmers' capacity in irrigation management, the farmers should have been trained well before the irrigation management transfer (Merrey *et al.*, 2007). However, this was not the case in Harran Plain, and farmers' training is still considered a 'to-do' after having practiced irrigated agriculture for more than a decade, and the 'blame-the-farmers' rhetoric of the governmental organizations remains valid.

According to the DSI officials, farmers are currently managing the irrigation systems themselves and irrigation associations have financial and administrative autonomy, which is considered as a big achievement of the irrigation management transfer. However, the farmers in Harran Plain disagree with this argument and continue to see the state as the ultimate decision maker and problem solver. Particularly, the smallholder and landless farmers have little faith in the capacity or willingness of irrigation associations to solve problems such as lack of canal maintenance and unequal distribution of water. On the other hand, many governmental and non-governmental actors consider the DSI as the 'dominant' and 'reputable' actor, both in teaching and imposing roles in the management of irrigation systems. This role of DSI constitutes an obstacle for entering into collaboration with other relevant actors, such as the Directorate General of Agrarian Reform, which carries out land consolidation and land distribution in many irrigated areas in Turkey including the Harran Plain.

Konya Closed Basin

Konya Closed Basin is a large-size basin (over 50,000 km^2) that encloses five provinces. The basin is a prominent agricultural area, where mainly irrigated and groundwater-dependent agriculture are practiced. Major agricultural products are water-thirsty plants such as sugar beet and maize, which are supported through price subsidies by the government. The on-farm irrigation method is primarily surface irrigation, whereas less than 5 per cent of the land is irrigated using the water-saving drip or sprinkle systems. The basin faces major water problems such as excessive water withdrawal and use, and the water users – especially the farmers in the region – bear the economic consequences.

The rate of groundwater withdrawal (over 4.5 billion m^3 per annum) in the basin is higher than that of replenishment (around 2 billion m^3 per annum), which causes a *c.*3 m annual drop in the groundwater table. Excess withdrawal primarily originates from the illegal wells drilled by farmers in their fields. DSI discloses that as high as 70,000 out of *c.*100,000 wells are illegal (not licensed by the DSI). As the groundwater retreats deeper each year, farmers encounter higher energy costs for water withdrawal, which in turn harms their products' competitiveness in the market and the sustainability of their agricultural income. Farmers typically refer to energy costs as one of the highest and continually increasing items in their cost structure. Groundwater in this basin is pumped from wells as deep as 100 m.

The environmentally unsustainable situation in the basin continues to attract the attention of NGOs who see farmers' irrigation habits and the irrigation infrastructure as the origins of the water problems. To help limit this they are running capacity-building and infrastructural projects in the area. Despite the ongoing efforts of NGOs, specifically the World Wide Fund for Nature (WWF) and the academic organizations in the region, there is still a lot to do to improve the sustainability of water resources in the basin.

Who is 'the public'?

Inhabitants of the basin are farmers who have similar preferences in terms of crop choice (sugar beet) and irrigation method (furrow irrigation). Several parties, including the local agricultural authority, the DSI, the agricultural engineering department of the local university, NGOs and the regional utility company, have expectations about farmers' behaviour, such as changing the way they practice agricultural activities. Their suggestions are centred on saving water by changing irrigation methods and crops. The public, often considered as being only the farmers and occasionally including the irrigation associations and cooperatives, is characterized as 'non-cooperative' because of their reluctance to follow suggestions related to water sustainability. Among the parties mentioned above, there is a common tendency to view the public as ignorant, and even unconscious about the requirements of their core business.

Similar to the Harran Plain case, irrigation associations manage the surface water and irrigation cooperatives manage the groundwater in Konya Closed Basin. How local authorities perceive irrigation associations and cooperatives varies and depends on their operations. Some irrigation associations and cooperatives are known for their historically embedded malpractices, which are often overlooked by the local authorities, mainly due to the lack of institutions and legislations that monitor and control the operations of these organizations. Nevertheless, there are some irrigation cooperatives known for their good practices in terms of the sustainability of the water and well-being of the community.

What is 'the participation'?

Farmers' participation in Konya Closed Basin is often used interchangeably with 'contribution' to the infrastructure costs and 'taking responsibility' for sustainability of water. The irrigation associations have been suffering from severe water loss in the open-top secondary and tertiary canals. The DSI proposed the renewal of the system with closed-section canals and covered half of the costs on behalf of the associations, provided credit with favourable terms for the remaining half and asked the associations to collect the credit from their members. In some associations, farmers were reluctant to pay the irrigation fee, both because their income is very limited and because they believed that the government should pay for this service upgrade. As one farmer stated: 'Everyone claims a part of my little income: the utility company, the irrigation association, the [agricultural production] cooperative… Do they want me to starve? It is I who should be supported'. The government regards these as operational costs to be borne by the farmer and thus does not sympathize with the previous types of statements. Similarly, irrigation associations are blamed for mismanagement by the farmers and by some other parties such as DSI and regional government agencies.

The responsibility for sustainable water use is a rather new demand that has been placed onto farmers. They are expected to follow suggestions such as shifting to drip irrigation, adjusting their crop selection for less water consuming types and not opening new irrigation wells. However, farmers see little incentive to follow these suggestions. An academic, experienced in some pilot projects in the region, puts it this way: 'It is hard to persuade them that they will get the same productivity with less water.'

Almost all parties agree that raising awareness among the farmers proved to be inefficient, and the best way to cooperate with them is through proper incentive and sanctioning mechanisms. Heavy sanctions however, will not always induce the desired cooperative behaviour. Regarding the continuously dropping groundwater table, opening new irrigation wells in the basin is forbidden yet farmers keep opening illegal wells in their farms. As the head of DSI regional office comments:

> It is illegal, yet I understand them… They need water to survive and the existing wells are not sufficient. What we need to do at this point is legalize the existing illegal wells and regulating their use through some quotas.

Why was participation introduced?

Participation is often confined to 'contribution to costs' by several parties. Yet, there is consensus on the idea that participation is regarded as the foremost requirement of water sustainability. The DSI, academia and NGOs agree upon the same game-theoretic proposition:

> If left alone, all farmers will act in their own interest and open new irrigation wells, spend more water than they need and deplete the groundwater faster. We need to convince them otherwise and make them aware of the greater good.

In the eyes of DSI, participation is also a means to engage farmers in the decisions:

> When asked to contribute financially, farmers are better at adopting the new systems. That is why we ask for a 50 per cent contribution from the farmers to the closed canal system infrastructure. This way, they will be more inclined to protect the close canal system.

Although participation practices are open to debate, some benefits are generally recognized. Academia and NGOs who run capacity-building and awareness-raising projects create quite an interest among the nearby farmers. Financial gains are still regarded as the most important outcome of practices that are being promoted in these projects. A farmer comments on this situation:

> Farmers in Altınekin [where a pilot project on drop irrigation is implemented by WWF] are getting the same crop as I do, but pay less for the water and electricity they consume to access to water. I want this project in my village, too.

Eastern Black Sea Region

The Eastern Black Sea Region has a basin area of 24,077 km^2 and is one of the richest basins when it comes to hydropower potential in Turkey. The basin has increasingly become the site of private hydropower development after the legal reforms of 2001. Due to its geographical and topographical advantages (high precipitation level, steep valleys), the region is among the most preferred places for the run-of-river type hydropower facilities. The estimated contribution of this region to the small-scale hydroelectric potential of Turkey is around 16 per cent (Uzlu *et al.*, 2011). As a result, at least 300 projects are licensed and in the process of construction. Despite its energy potential, the basin also hosts many species as well as 2.5 million people, many of whose livelihoods depend on these rivers.

This small-scale hydropower development is regulated by 'water-use right contracts' that are prepared by the DSI. Accordingly, water-use rights to produce electricity are granted to companies for only forty-nine years. Since these contracts allow the privatization of water-use rights, communities now have limited or no rights to use the river for other purposes such as their domestic and agricultural use. As emphasized in the contracts, the ultimate rights of rivers belong to the state, and private companies obtain user rights. However, the company is responsible for the construction, management, operation and security of facilities and the potential risks derived from the usage of rivers. Because rights and responsibilities over the use of rivers are transferred to the private entities, there are serious uprisings from the rural communities in the region regarding their exclusion both in the current legal framework and in the decision-making processes.

Who is 'the public'?

Stakeholders from the public and the private sector can participate in the decision-making process through the EIA notification meetings. According to the Ministry of Environment, the transparent and democratic participation of different stakeholders is a necessary condition in the EIA process, since it is a process for finding out and preventing the future social and environmental implications of development programmes (Ahmad and Wood, 2002). The term 'public' is used to allow everybody to participate in the EIA legislation of 2008 (Scheumann *et al.*, 2011). The private sector and affected communities have various views on the effectiveness of EIA meetings. Some of the locals from the region argue that meetings are far from being inclusive and democratic in practice. The majority of the communities do not participate in these meetings because they believe their views are being ignored and undermined. Private sector are willing to organize these meetings however some perceived them as loss of time and money.

What is 'the participation'?

The EIA process and thus the participation of the public only enter the project processes at the later stages of development (Scheumann *et al.*, 2011). As a part of this process, EIA meetings are designed for consultation and sharing of information but the outcomes barely influence the decision making. Especially in the case of the Eastern Black Sea Region, public participation is very limited due to the high level of distrust between the local people and private companies. Some villagers choose not to recognize the process because they feel alienated from the process. Local knowledge and values are not recognized and excluded from the decision-making process.

Social and environmental concerns led to the mobilization of a powerful opposition towards hydroelectricity projects in the region. Apart from the activist groups in the Mediterranean Region (southern Turkey), such as Save Alakir Network and Yuvarlakcay Protection Platform, many other activist groups from the Black Sea region like Karadeniz Resurrection, Brotherhood of Rivers Platform and Ikizdere Association protest against hydropower projects during the EIA meetings. In many of these cases, non-participation has also become a position since some groups argue that participation would imply an indirect support of the process.

Why was participation introduced?

The international regulations and EU approximation process have influenced the EIA processes in Turkey. Recent changes in the EIA regulation have put more emphasis on public participation. Accordingly, stakeholders have to be informed and consulted about projects during the planning process. However, there is no consensus on the rationale behind introducing public participation into the planning of hydropower projects. According to the Ministry of Environment and Urbanization, public participation is necessary to build trust between communities and project owners and to provide a more transparent process by influencing the EIA process before projects are approved. This position is in line with the argument that the inclusion of public opinions in the EIA can lead to the development of a sense of ownership and reduce conflict (Shepherd and Bowler, 1997). However, a DSI official explains: 'The state is the ultimate owner of these rivers, we can consult people but at the end we are making the final decision'.

Although controversial in terms of inclusiveness, EIA notification meetings are the only platform bringing the public and the private sectors together. Through these meetings, people get information and express their opinions about the projects. However, due to the distrust between stakeholders, most often these meetings do not create a platform for a dialogue. On one hand, the majority of the private sector perceives EIA notification meetings as ineffective and time-consuming and, thus, not cost effective. On the other hand, some villagers perceived that the EIAs 'serve for the advertisement of their company'.

In the Eastern Black Sea Region, the active resistance groups against hydropower projects perceive EIA meetings as a mechanism that perpetuates the privatized water development regardless of the opposite views of the local people. However, some companies do recognize the importance of people's consent as we see in the Yuvarlakcay case in south-western Turkey, where a company has withdrawn its project as a result of the local resistance (Islar, 2010).

8.5 Discussion: the contested nature and the transfer of public participation

The analysis of discursive positions in three cases of public participation in Turkish water management reveals both similarities and differences regarding the contested nature of public participation. These similarities and differences can be attributed to various contextual factors as well as the characteristics of the actors.

Constituting typical examples of participatory irrigation management, the Harran Plain and the Konya Closed Basin cases provide insights about the interactions among the water users (individual farmers and WUOs) and the DSI. The discursive disagreements and differences among the water users and the DSI can be attributed to the imposition of the participatory approach on farmers without making sure that the farmers possess the required individual and organizational capacities for irrigation management so that they could share similar meanings of 'participation'. Furthermore, the scope of the 'public' is often limited to the local elite and the powerful farmers that have easier access to physical and financial resources. This power asymmetry is, in most situations, taken for granted by both the farmers and the DSI although it implies an 'elite capture', which can influence the decisions and inhibit the enforcement of sanctions (Kadirbeyoğlu and Özertan, 2011).

In the Eastern Black Sea Region case, participatory mechanisms are incorporated into the processes of hydropower planning. This case complements the other cases in terms of both the sectoral use of water resources and the phase of water management. Similar to the other cases, public participation is contested due to the divergent uses attributed to 'participation'. While the DSI and the private sector perceive the involvement of the public as a way to share information and consultation, local communities demand a more active role in the decision-making process. The Eastern Black Sea Region differs from the other two cases in two aspects. First, the affected local communities mobilize against the mechanisms that exclude their effective participation and, thus, aim to extend the scope of 'public'. Second, the role of DSI is less prominent in the participatory process since the risks and responsibilities of hydropower development are partly transferred to private actors (the companies taking over the use right of water resources). Nevertheless, in all the cases, the DSI holds the property rights for water. This fact positions the DSI as a prominent actor in the governance system as well as the ultimate decision-maker. The scope and the forms of participation by other parties, including public members and private actors, are shaped by the discourse that the DSI leads.

The contextual factors and the characteristics of actors influence the implications of the contested nature of public participation. The long-standing hierarchical structure in the Harran Plain manifests itself in all types of decision-making processes and limits the scope of the public to a privileged group. Furthermore, the lack of previous experience in irrigation management and the abrupt devolution of responsibility to the newly established irrigation associations are key determinants of the ongoing discourse on the incapability of farmers and their organizations in really participating in participatory irrigation management. Similarly, in the Konya Closed Basin case, actors' capacity play a prominent role in the way participation is interpreted and practised. References related to the incapability of farmers were often heard, which might explain why the discursive positions about participation are confined to contributing to costs and following the suggestions provided by the more capable actors, neither of which is a genuine form of participation. Resistance movements in the Eastern Black Sea Region case show that effective and inclusive participation is a necessary condition for both the future of hydropower development projects and social and political sustainability. Many projects cannot even be constructed due to the active resistance that eventually requires additional money, labour and time from the private companies. According to a lawyer from the Turkish Water Assembly, around 100 (almost half of the total proposed) projects are on hold due to court decisions (Y. Okumusoglu, personal communication, 2012).

8.6 Conclusions

In this chapter, we elaborated on the underlying mechanisms of the contested nature of public participation in water management and the implications of this contested nature on its transfer to non-participatory settings. Our empirical findings indicate that public participation becomes contested when actors with divergent characteristics attribute different definitions and uses to the concept. Combined with the influence of contextual factors, these divergent characteristics lead to several implications about the transfer of public participation to top-down and hierarchical water management settings. The immaturity regarding the adoption of participatory mechanisms in the broader arenas of environmental decision-making are also reflected in the water sector.

At the case-specific level, the low degree of individual and institutional capacity to engage proactively in the decision-making processes constitutes an obstacle in clarifying and expanding the boundaries of public participation. Despite the possibility of involving a broader scope of participants in multiple phases of decision-making processes, the more powerful actors continue to shape the dominant definition of public participation. Therefore, introducing participatory mechanisms that have been developed by external actors such as the EU and the World Bank remains insufficient for overcoming the power asymmetries and hierarchical structures at the local level.

The effectiveness of the transfer of participatory mechanisms to top-down and hierarchical water management settings can be improved in several ways. Regarding the implementation of participatory irrigation management, ensuring the involvement of disadvantaged farmers, primarily the landless and smallholder farmers, is of utmost importance. Furthermore, when the managerial and financial performance of WUOs are in question, the monitoring and evaluation by third-party (or independent) organizations can be useful to change the perception of farmers and other actors about the functioning of the WUOs. For the participatory planning of hydropower development, an independent stakeholder advisory group, which can voice local concerns, provide local knowledge, help to identify key issues and facilitate cooperation between different groups, can contribute to the public participation processes.

References

Ahmad, B. and Wood, C. (2002) 'A comparative evaluation of the EIA systems in Egypt, Turkey and Tunisia', *Environmental Impact Assessment Review*, vol. 22, no. 3, pp. 213–234.

Aküzüm, T., Kodal, S. and Çakmak, B. (1997) 'Irrigation management in GAP', *Water Resources Development*, vol. 13, no. 4, pp. 547–560.

Arnstein, S.R. (1969) 'A ladder of citizen participation', *Journal of the American Institute of Planners*, vol. 35, no. 4, pp. 216–224.

Bressers, H. and Lulofs, K. (eds) (2010) *Governance and Complexity in Water Management: Creating Cooperation through Boundary Spanning Strategies*, Edward Elgar, Cheltenham/Northampton, MA.

Creighton, J.L. (2005) 'What water managers need to know about public participation: one US practitioner's perspective', *Water Policy*, vol. 7, no. 3, pp. 269–278.

Day, D. (1997) 'Citizen participation in the planning process: An essentially contested concept?', *Journal of Planning Literature*, vol. 11, no. 3, pp. 421–434.

DSI (Devlet Su İşleri) (2011) *Faaliyet Raporu 2010 (Annual Report 2010)*, DSI, Ankara.

DSI (Devlet Su İşleri) (2012) 'Hizmet alanları: tarım (Service fields: agriculture)', online, available at: www.dsi.gov.tr/hizmet-alanlari/tarim, accessed 5 November 2012.

Gallie, W.B. (1956) 'Essentially contested concepts', *Proceedings of the Aristotelian Society New Series*, vol. 56, pp. 167–198.

Grontmij (2005) *Draft River Basin Management Plan Büyük Menderes and Gediz, Implementation of the Water Framework Directive in Turkey, Final Report*, Grontmij Consulting Engineers, Houten.

Hajer, M. and Versteeg, W. (2005) 'A decade of discourse analysis of environmental politics: Achievements, challenges, perspectives', *Journal of Environmental Policy and Planning*, vol. 7, no. 3, pp. 175–184.

Innes, J.E. and Booher, D.E. (2004) 'Reframing public participation: Strategies for the 21st century', *Planning Theory and Practice*, vol. 5, no. 4, pp. 419–436.

Islar, M. (2010) 'Privatized rivers of Turkey: The case of Yuvarlakcay', paper presented at the Berlin Conference on Human Dimensions of Global Environmental Change, 8–9 October 2010, Berlin.

Kadirbeyoğlu, Z. (2008) 'Decentralization and democratization: The case of water user associations in Turkey', PhD thesis, McGill University, Montreal.

Kadirbeyoğlu, Z. and Özertan, G. (2011) 'Users' perceptions of water user associations: Evidence from three cases in Turkey', Boğaziçi University Department of Economics, Working Paper 2011-1, pp. 1–29.

Kibaroglu, A. and Baskan, A. (2011), 'Turkey's water policy framework', in A. Kibaroglu, W. Scheumann and A. Kramer (eds) *Turkey's Water Policy: National Frameworks and International Cooperation*, Springer, Heidelberg/Dordrecht/ London/New York.

Kibaroglu, A., Baskan, A. and Alp, S. (2009) 'Neo-Liberal transitions in hydropower and irrigation water management in Turkey: Main actors and opposition groups', in D. Huitema and S. Meijerink (eds) *Water Policy Entrepreneurs: A Research Companion to Water Transitions Around the Globe*, Edward Elgar, Cheltenham.

Merrey, D.J., Meinzen-Dick, R. and Mollinga, P. (2007) 'Policy and institutional reform: The art of the possible', in D. Molden (ed.) *Water for Food – Water for Life, Comprehensive Assessment of Water Management in Agriculture*, EarthScan, London.

Molle, F. (2008) 'Nirvana concepts, narratives and policy models: Insights from the water sector', *Water Alternatives*, vol. 1, no. 1, pp. 131–156.

Mostert, E. (2003) 'The challenge of public participation', *Water Policy*, vol. 5, no. 2, pp. 179–197.

Özerol, G. (2013) 'Institutions of participation and environmental sustainability: A multi-level analysis from irrigation management in Harran Plain, Turkey', *International Journal of the Commons* vol. 7, no. 1, pp. 73–91.

Özerol, G. and Newig, J. (2008) 'Evaluating the success of public participation in water resources management – five key constituents', *Water Policy*, vol. 10, no. 6, pp. 639–655.Phillips, N. and Hardy, C. (2002) *Discourse Analysis: Investigating Processes of Social Construction*, Sage, Thousand Oaks, CA.

Scheumann, W., Baumann, V., Mueller, A.L., Mutschler, D., Steiner, S. and Walenta, T. (2011) 'Environmental impact assessment in Turkish dam planning', in A. Kibaroglu, W. Scheumann and A. Kramer (eds) *Turkey's Water Policy: National Frameworks and International Cooperation*, Springer, Heidelberg/Dordrecht/ London/New York.

Shepherd, A. and Bowler, C. (1997) 'Beyond the requirements: Improving public participation in EIA', *Journal of Environmental Planning and Management*, vol. 40, no. 6, pp. 725–738.

Sneddon, C. and Fox, C. (2007) 'Power, development, and institutional change: Participatory governance in the Lower Mekong Basin', *World Development*, vol. 35, no. 12, pp. 2161–2181.

SPO (State Planning Office) (2006) 'Ninth Development Plan (2007–2013)', SPO, Ankara.

SPO (State Planning Office) (2007a) 'Dokuzuncu Kalkınma Planı: Çevre Özel İhtisas Komisyonu Raporu (Ninth Development Plan: Report of the Specialization Commission on Environment)', SPO, Ankara.

SPO (State Planning Office) (2007b) 'Dokuzuncu Kalkınma Planı: Toprak ve Su Kaynaklarının Kullanımı ve Yönetimi Özel İhtisas Komisyonu Raporu (Ninth Development Plan: Report of the Specialization Commission on the Use and Management of Soil and Water Resources)', SPO, Ankara.

UNECE (United Nations Economic Commission for Europe) (1998) Convention on Access to Information, Public Participation in Decision-making and Access to Justice in Environmental Matters, UNECE, Aarhus.

Uzlu, E., Akpinar, A. and Komurcu, M.I. (2011) 'Restructuring of Turkey's electricity market and the share of hydropower energy: The case of the Eastern Black Sea Basin', *Renewable Energy*, vol. 36, no. 2, pp. 676–688.

Vermillion, D.L. (1997) *Impacts of Irrigation Management Transfer: A Review of the Evidence*, Research Report 11, International Irrigation Management Institute, Colombo.

9 Multi-stakeholder partnerships in fragile political contexts

Experiences from the Palestinian water and waste sector

Irna van der Molen and Nora Stel

9.1 Introduction

The concept of multi-stakeholder partnerships (MSPs) is used as both an analytical tool and a policy prescription – often at the same time. On the one hand, a vast academic literature is dedicated to MSPs, drawing on the theoretical fields of governance, networks and management (Noor, 2010).[1] On the other hand, the MSP notion has emerged from development practitioners' and policymakers' conceptions of dialogue and partnership.

We define MSPs as cooperation between different stakeholders that have an interest in a problem and together engage in a process of dialogue and cooperative action to address this problem (Vermeulen *et al.*, 2008, p. 97; Hemmati, 2002, p. 2). These 'different stakeholders' are often conceived as coming from the private, public and civil sectors. Thus, MSPs can be seen as extended versions of, or related to, public–private partnerships (PPPs), which are defined as 'arrangements whereby private parties participate in, or provide support for, the provision of infrastructure, resulting in a contract for a private entity to deliver public infrastructure-based service' (Grimsey and Lewis, 2004 cited in Multipart, 2008, p. 43).

The common ground for an interest in MSPs, among development practitioners and policymakers, can be found across the processes of interaction and governance (Noor, 2010). As Nuijten eloquently states: 'the MSP concept is buttressed by a *belief* in the power of dialogue and consensus-building in breaking down institutional and power barriers, and the ability of people at the local level to manage local resources' (Nuijten, 2004, p. 116, original emphasis). The leading international donors and policymaking institutions propagate this belief in MSPs in water management in general, and in the water supply, sanitation and hygiene (WASH) sector in particular.

For example, the United Kingdom Department for International Development (DfID, 2011, p. 41) welcomes MSPs as a 'promising mechanism towards building cooperative water management' and calls the water crisis a 'crisis of governance' (Plummer and Slaymaker, 2007). The United Nations sees multi-stakeholder governance as a pillar of WASH policy (United Nations Development Programme, 2004). The Organisation for

Economic Co-operation and Development (OECD), too, considers partnerships and 'multi-stakeholder dialogue' as the solution for lacking state capacity and political will (2008, p. 34, 2012, p. 19).

Motivations for MSP dissemination invariably – either implicitly or explicitly – refer to the ability of MSPs to operationalize integrated water resource management, good governance and stakeholder participation, which are among the key concepts that dominate the debates on the management of water and waste over the last decades (see also Chapter 2). In this sense, the MSP concept is a rather complex one that contains elements of all three panaceas or perceived universal remedies that are described in Chapter 2. MSPs are perceived as a way to enhance equity, accountability, democracy, transparency, participation and integration (Noor, 2010, p. 36). MSPs are praised for their ability to link top-down and bottom-up; indigenous and foreign; and state and non-state approaches and are assumed to make service delivery more inclusive, participatory and effective (Hofmann, 2009, cited in Noor, 2010, p. 36; van Tongeren and van Empel, 2007, p. 14).

With the adoption of Agenda 21, specifically its section on 'strengthening the role of major groups', at the Earth Summit in Rio de Janeiro, partnerships and MSPs were introduced as new forms of governance. The importance of partnerships was reiterated at the Johannesburg Summit in 2002 (Multipart, 2008). Recognizing that neither the state, nor the market, nor communities seem to be able to manage water on their own, MSPs were coupled with integrated water resource management (De Boer and Pfisterer, 2009). The emergence of MSPs can therefore also be perceived as part of a process of diversification of governance modalities (Sørensen and Torfing, 2005, p. 199; Pierre and Peters, 2000).

In this chapter, we discuss both the transfer of the MSP concept in the water and the waste sector, and the underlying assumption that this governance concept is – with some adjustments – applicable and effective also in fragile contexts that are characterized by conflict, political instability and institutional multiplicity.

Based on fieldwork in the Palestinian Territories, we suggest that multi-stakeholder governance of water, sanitation and waste management is a requirement for successful service provision in these sectors because of the fragility of the governance context. This is already widely recognized and applied by relevant Palestinian stakeholders and policymakers. As such, it is misguided to assume that MSPs are an entirely new construct, a tool that needs to be introduced to 'modernize' water management in developing countries and/or fragile situations. Our cases show that MSPs are not a new concept introduced by donor organizations, but emerge both with and without donor pressure. The relevant questions, then, evolve not around whether to promote or transfer MSPs, but how to enhance their effectiveness and added value in fragile settings, through the transfer of knowledge. The research question that guides our analysis is:

> *How applicable are MSPs in fragile contexts? How can donor organizations play a role in enhancing the effectiveness and added value of MSPs in fragile situations, if at all?*

To answer this question, we first describe the MSPs and their features. This includes the characteristics of each MSP (initiation and origin, objectives, actors, activities and approach, funding); the governance of each MSP (preparation phase, decision making and implementation, information and communication, completion, continuation; and the output of each MSP (achievement of objectives and sustainability).

Second, we analyse the added value and effectiveness of MSPs in the WASH sector. Effectiveness is defined here as the achievement of the objectives as set out by the stakeholders and the sustainability of these achievements. In addition to three cases from the water sector, we include a case from the waste sector. Due to its impact on the quality of the groundwater, we consider adequate solid waste management 'part and parcel' of effective water resource management in a region that is dependent on groundwater. We refer to this combined water and waste sector as WASH+.

Third, we focus on the contextual factors affecting the governance and effectiveness of the MSPs studied. Throughout the chapter we refer to several ideas of the Contextual Interaction Theory (CIT). This is relevant since the core assumption underlying CIT (that contextual factors assert important influence over the process and effectiveness of policy implementation) corresponds well with our main arguments. We do not, however, claim to apply all components of CIT in a comprehensive way. We particularly draw on the distinction between wider, structural and case-specific contexts to address the impact of political instability (as part of the wider context) and institutional multiplicity (as part of the structural context) on MSPs. With *structural context*, we imply the multiplicity of governance levels, networks, perspectives, strategies and resources as envisioned by Bressers and de Boer in Chapter 3.

Two contextual factors are selected to illustrate the challenges of multi-stakeholder governance concepts in fragile settings: the Israeli–Palestinian conflict at large; and the fragmentation of the West Bank in areas A, B and C and the parallel institutions resulting from the governmental Hamas–Fatah divide.

9.2 Methodology

Researchers from the Palestinian Hydrology Group (PHG), together with the authors, collected the data between 2009 and 2011 in a research project on four MSP cases in the water, sanitation and solid waste sectors. The research was part of the Peace, Security and Development Network (PSDN) initiated by the Dutch Ministry of Foreign Affairs (Stel *et al.*, 2012). The qualitative methodology comprised extensive literature and secondary sources research; in-depth interviews and focus groups with a total of ninety-eight respondents; and four validation and consultation workshops with respondents and experts.[2] The analysis of the MSPs was facilitated by a theoretical and analytical framework, which was developed by the consortium (Noor *et al.*, 2010). The validation and consultation workshops that took place mid-term and at the end of the project further contributed to adequate interpretation and analysis of the data.

In the Palestinian Territories, the cases were selected on the basis of the general criteria from the PSDN. The selection criteria were that the MSP is organized around utilities such as water supply and sanitation, wastewater, and solid waste management; involves private, public and civil actors; and has a measurable output. An additional desire of local partners was to include initiatives which were of strategic importance or unique in some way. These unique features are elaborated upon in Section 9.4. Research on the case study in the Gaza Strip was hampered by impeded access to the Gaza Strip and lack of possibilities from PHG research staff in Gaza to communicate with the Head Office in Ramallah: video and Skype connections failed consistently. In the next section, we first discuss the development of MSPs as a tool in donor policy in the section below.

9.3 The features of Palestinian MSPs

The first case study concerns the rehabilitation of groundwater wells around two cities in the Northern West Bank: Qalqiliya and Tulkarem. Both cities are located close to the Green Line[3] and the Separation Wall, the barrier built by Israel on the Palestinian side of the Green Line. The problems the MSP sought to address through the well rehabilitation included:

1 frequent interruption of water supply;
2 excessive operation costs for the wells;
3 high debts among well owners and municipalities running the wells;
4 limited well capacity; contamination by sewage water and flooding from settlements; limited pump capacity;
5 insufficient water supply to cover the needs of communities;
6 decreased agricultural productivity due to abandoned land; and
7 lack of skills among well owners and operators.

Farmers, groundwater well owners, village councils and municipalities, national governmental organizations, non-governmental organizations (NGOs), and international organizations were able to ensure good cooperation and results by substantially increasing access to water[4] for domestic and agricultural use.

The second case study regards a landfill located in the Zahrat Al Finjan region approximately 17 km south-west of the Northern West Bank city

Jenin. Before the establishment of the landfill, poor management of the solid waste collection and disposal services in Jenin governorate were considered to affect soil, public health, environment and water resources. The pollution created by the former unregulated waste dumps that have affected the groundwater supplies through infiltration in the soil was a great concern. This was not only an issue for the Palestinian authorities, but also for the Israeli authorities, since the surrounding Israeli settlements were extracting their water from the same source. The Zahrat Al Finjan landfill was initiated in 1998 by the World Bank under the Solid Waste and Environmental Management Project. The project aimed to implement an environmentally sound solid waste management system for the Jenin governorate, that included:

1 safe disposal of solid waste in controlled landfills;
2 environmentally sound collection and transfer of solid wastes to landfills in Jenin district;
3 stakeholder ownership of the project;
4 enhanced implementation capacity; and
5 enhanced monitoring and enforcement capacity.

The World Bank worked together with the European Community Humanitarian Office (ECHO), a number of ministries, local governmental organizations, such as the Jenin Service Council; the Project Implementation Unit and Jenin governor; Israeli authorities, and contractors. The success of this landfill project is remarkable, since the Second Intifada started almost immediately after the start of the project. Furthermore, the Israeli Civilian Administration granted approval to construct this facility while many other proposals were rejected or delayed. As a result, it is one of the few regulated and formal landfills in Palestine.

The third case study is a wastewater treatment plant located in Kharas in the southern part of West Bank, approximately eight kilometres west of Hebron City. The treatment plant project was a response to the local community that sought an alternative for sewage disposal into cesspits that polluted their groundwater as well as their drinking water supply. Local stakeholders (inhabitants, landowners, farmers, local councils) worked together with governmental organizations (both at local and national level), NGOs and international organizations. The project was among the first wastewater treatment plants implemented in the rural areas of the southern part of the West Bank. It is innovative since it collects and treats sewage at the village level, rather than at the household level. It introduced low-cost and efficient technology for sewage treatment with the objectives to improve hygienic conditions; protect water quality; reduce pollution loads; and demonstrate possibilities for sound collection and treatment of wastewater at the village level. During the first seven years, the treatment facility offered a good sanitary solution. Thereafter, the operation failed due to challenges in management capacity, communication and revenue collection.

The fourth case study concerns a desalination plant in the Gaza Strip's Bureij camp. This station was established after a 2006 assessment showed how much citizens in Bureij area suffered from high salinity of drinking water. The situation in Gaza is not only constrained by repeated episodes of violence and military operations and by withdrawal of donor organizations after Hamas won the elections, but also by a double public administration (one run by Hamas, the other one inactive under the Palestinian Authority). The MSP's activities included rehabilitation and upgrading of a municipal well and installation of a desalination plant; construction of a reservoir to blend the water with water from other wells; closing of two contaminated wells; and the provision of training for the operation and management of the water supply system. The PHG and the Italian Civil Group of Volunteers (*Groppo Volenterinto Civile*) were responsible for implementation. They worked together with Terra des Hommes Italy, a local water committee, the Palestinian Water Authority, the Municipality of Bureij, the Coastal Municipal Water Utility, and Gaza Electrical Distribution Company.

9.4 Added value and effectiveness of the Palestinian MSPs

In the four Palestinian MSPs studied, the MSP approach was perceived as a self-evident and inevitable way of resolving problems in the water, wastewater and waste sector, rather than as a conscious or innovative way of arranging WASH+ services. Our research suggests that MSPs exist in Palestine as an empirical reality, but have little significance as an explicit policy concept. In Palestine, MSPs are implemented on a case-by-case basis, dependent on individual donors and NGOs. Those MSPs predominantly originated as organically emerging cooperation mechanisms rather than as results of conscious policymaking:

- In case 1, the MSP in Qalqiliya and Tulkarem, a group of farmers sent applications to the village council to help them rehabilitate some of their irrigation wells. The village council contacted the PHG to assist them for the rehabilitation of these wells and the PHG subsequently wrote a proposal and submitted this to the European Union and invited farmers and well owners to discuss their problems.
- In case 2, the Zahrat Al Finjan landfill, the MSP was designed explicitly as a MSP under the World Solid Waste and Environmental Management Project. Several workshops and meetings with key experts, however, indicated that such predetermined 'MSP-ness' is not common in Palestine.
- The MSP in Kharas, case 3, also emerged without any MSP blueprint or ideology. It was a response from the PHG to local community needs.

- The MSP in Al-Bureij, case 4, started as a partnership between the Civil Group of Volunteers and the PHG. They cooperated with local community-based organizations, the municipality, the Palestinian Water Authority (PWA), and broadened partnership with participation of semi-public entities (utilities) to achieve the objectives.

In terms of the governance and implementation of the MSPs, our cases show that MSPs can have a multi-stakeholder governance structure or practice without being officially initiated as an MSP. These MSPs are therefore the result of the specific history consisting of previous decision-making processes (see also Chapter 3). A history of professional cooperation and joint decision-making was found to contribute positively to the motivation of actors, to the functioning of MSPs and thereby to their effectiveness, but only when this history of cooperation was perceived as unproblematic. Such organically emerging processes are not necessarily ungoverned, anarchical or homogenous, but can arrive at much the same practices as 'official' MSPs (Warner cited in Nuijten, 2004, p. 118).

- The tasks of the main actors in case 1 had been specified pragmatically and a steering committee was eventually established (ECHO, 2010, p. 64–66). The good cooperation between stakeholders was, according to our respondents, the result of a long history of professional cooperation and an intrinsic appreciation of cooperation by beneficiaries and partners. However, the sustainability of the project could not be ensured, despite being a priority among MSP members from the start.
- Case 2 seems to be exceptional as it explicitly opted for an MSP approach from its initiation. Respondents and evaluation reports recall intensive and institutionalized communication and dialogue among all stakeholders involved throughout all phases and continuous investment in realizing both anonymity and equity of all partners in the MSP's decision-making process (World Bank, 2009). The World Bank's on-the-ground-support unit provided for effective communication with the Israeli civilian and military authorities.
- The governance context of case 3 yields a less positive picture – of problematic communal relationships and a lack of internal capacity hampering communication and effectiveness. While the governance of the MSP was characterized by multi-stakeholder interaction, it was not a smooth process. Lengthy and intensive processes of negotiation were required among and with stakeholders; between village council and landowners; between community and ministries; and between the PHG and the national governmental institutions involved (EHP, 2002).
- Due to political developments (electoral victory of Hamas) and the subsequent change of government in Gaza, the membership of stakeholder organizations in case 4 changed repeatedly. The uncertainty

during transition has affected not only the local water committee, but also the Municipality of Al-Bureij and the replacement of staff at PWA in Gaza.

When it comes to the effectiveness and output of the MSPs studied, several observations can be made. As noted above, multi-stakeholder involvement in the governance and implementation of a project does not guarantee achievement of all objectives as was seen in cases 3 and 4. The MSP established for the Zahrat Al Finjan landfill, which was most consciously and formally conceived of as an MSP, did have the most positive and sustainable output according to respondents. Whether this was due to the formalization of cooperation within the project or other factors such as the more extensive donor commitment is hard to establish. Nonetheless, respondents were convinced that the inclusive and communicative nature of the project contributed to its success. For the other three cases – MSPs in practice but not on official paper – achieving objectives in terms of 'hardware' (construction) proved significantly more feasible than improving 'software' (institutional strengthening, capacity building, awareness raising).

Considering the above, our findings are multiple – and perhaps paradoxical. On the one hand, the WASH+ initiatives we studied are functioning as MSPs even if they are not instigated as MSPs or officially presented and recognized as such. MSPs are thus an empirical reality even without being influenced by policy prescriptions. At the same time, we note that the output of the one MSP that was formally initiated and governed as an MSP (case 2) was more successful and sustainable in output terms. Although stakeholders attributed the success of this project to a range of factors, stakeholders repeatedly commended the MSP structure and governance of the Jenin Joint Service Council. Finally, for the effectiveness and outputs of the MSP, the added value of active involvement and commitment by donor organizations in MSPs is particularly recognized by stakeholders, primarily in relation to pressure on Israeli authorities for granting permits required during project implementation.

9.5 The influence of the wider context on Palestinian WASH+ MSPs

It is important to note that not everyone shares the widespread policy enthusiasm for MSPs in supranational policymaking networks described in Section 9.3. The 'neutral' management and government discourse in which MSPs are often cast, risks depoliticizing processes that are in fact highly political, and essentially determining 'who gets what water, when and how' (Tropp, 2005). Moreover, the difficulty of generating a common vision and action plan in conflict-affected situations is underrepresented in MSP research (Warner, 2006, p. 20). This multifaceted critique on MSPs as a transferable policy concept is even more relevant in so-called fragile situations.

Two core features characterize politically fragile or hybrid situations and set them apart from 'traditional' developing settings: (1) (the risk of recurrent) violent conflict and the vulnerability to external shocks related to it; and (2) politicization of public goods provision and institutional multiplicity (Naudé *et al.*, 2011; Boege *et al.*, 2009; Hofmann, 2009).[5] Below we explore the influence of a wider fragile context on the structural and case-specific contexts of our cases. For the Palestinian wider context, we draw attention to three concurrent attributes signifying fragility. The first contextual attribute is the Israeli-Palestinian conflict at large, related to the risk of (recurrent) violent conflict. The second is the institutional multiplicity as exemplified by the governmental divide between Fatah and Hamas in the Gaza Strip. Third, there is the administrative fragmentation in areas A, B and C in the West Bank. We therefore focus on factors that are related to the political context, and much less to the economic, cultural or technological contexts.

The impact of the Israeli–Palestinian conflict on the institutional setting

The present problems regarding water resources management in Israel and the Palestinian Territories have their roots in the Partition Plan of 1947; the unilateral declaration of the establishment of Israel in 1948; the 1967 war; the Oslo Agreements and internal developments. Israel took control over the water resources in the West Bank after the 1967 war. This was arranged through a number of Military Orders. Before the Oslo Agreements, the West Bank Water Department was the largest water distributor in the West Bank. The Interim Agreement of 1995 facilitated the transfer of responsibilities to the PWA. This transfer has resulted in an intricate and multilevel governance system characterized by the inherent (even inevitable) involvement of multiple stakeholders.

First, the PWA and the Israeli Water Authority are legally obliged to work together in the Joint Water Committee, which was established as a result of the Interim Agreement. Although Joint Water Committee decision making is officially based on consensus and both countries are formally equally represented, stakeholders perceive that the committee reiterates existing power imbalances between Israel and Palestine.[6] A PWA spokesperson called it 'the unwelcome stakeholder' (Stel, 2012, p. 5).

Second, cooperation between the PWA and municipalities is institutionalized and arranged by Local Authorities Law no. 1, 1997. However, the lack of sovereignty of the Palestinian authorities, and the subsequent lack of trust in the authorities, has led to a situation in which citizens do not rely on the state to provide certain basic services. As a result, some community-based organizations and NGOs are trying to improve the local situation with regard to water supply, sanitation and hygiene, in cooperation with local authorities. In this regard, various private enterprises, ranging from local truck vendors to fully fledged PPPs, fill the gap left by the public sector by responding to citizens' demands and making profit from the situation.

Third, the Palestinian WASH+ sector – as in most other developing countries – includes many national and international NGOs and donor agencies. Bilateral and multilateral donors are involved in a variety of coordination mechanisms in various sectors. International NGOs and donors that are involved in water and sanitation coordinate their efforts in the Association of International Development Agencies; the Emergency Water, Sanitation and Hygiene Group (EWASH); the Consolidated Appeal Process for the Occupied Palestinian Territories and the Organization for Coordination of Humanitarian Affairs.

The uncertainty, politicization of service delivery and institutional and practical restrictions in terms of permits, mobility and access to water resources brought about by the conflict create a greater need for actors to work together. In our case studies, we found several examples of how this context shaped the motives, cognitions and resources of the actors involved in the MSPs. Regarding case 1, the impact of wider context specifically influenced first the *cognitions* and subsequently the *motives* of local authorities. In 2003, a group of farmers from the Qalqiliya and Tulkarem area sent applications to the village council asking for urgent intervention to help them rehabilitate some of their irrigation wells to use them again for irrigation purposes. Initially, the rehabilitation of irrigation wells did not have a high priority for the village council. This changed, however, after the Separation Wall was built close to Qalqiliya and Tulkarem.[7] The village council then contacted the PHG to assist them in the rehabilitation of these wells which eventually resulted in the 'Urgent Action to secure access of the population in the Occupied Palestinian Territories' project funded by ECHO.

The wider conflict context first changed local authorities' perceptions, or *cognitions*, of the problem at hand: from an agricultural predicament to a political emergency. While the rehabilitation of irrigation wells initially did not have a high priority for the village council, this changed after the Separation Wall was built in Qalqiliya and Tulkarem in 2002. Based on Israeli law; farmers risk losing their land if they do not cultivate it, including Palestinian farmers with land east of the Green Line. As a result, the motives of key decision-makers shifted accordingly. Interests that were initially of a practical nature (access to water, improvement of farmers' livelihoods) were now coinciding with strategic interests (continuation of cultivation of Palestinian lands between the Green Line and the Wall to maintain Palestinian land).[8] This conflict-cognizant approach also manifested itself in the process of planning and implementation of the MSP in which the potential risks related to the conflict were carefully assessed and accommodated. These risks included delays in receiving permits, a 'stop of works' order by the Israeli Defence Forces and military operations. Great effort was taken to minimize risks that water users would be cut off from water and to obtain the *resources* (in this case permits) for the transfer of equipment required for rehabilitation of the wells.

Case 2 was affected by the wider conflict context as well. The start of the landfill project in October 2000 coincided with the start of the Second Intifada, which threatened to crucially undermine the available *resources* to initiate this ambitious project. Not surprisingly, some international staff decided to leave the country. Furthermore, the situation created great reluctance among international contractors and consultants to engage in bidding activities. The hostilities, that started in September 2000 and rapidly gained momentum between September and December, had a major impact on the project in terms of increased requirements by the Israeli authorities for constructors to apply for permits to use and transport certain types of equipment. The conflict dynamics caused delays in preparation and execution of the core landfill component due to the closing-off of Jenin by the Israeli Defence Forces and because 'project offices were shot at, waste containers and access roads were damaged, and it was difficult for project staff, consultants and the [World] Bank supervision to operate' (World Bank, 2009, p. 5).[9]

The impact of the government divide between Hamas and Fatah

In January 2006, Hamas won the Palestinian parliamentary elections. This resulted in political violence between Fatah and Hamas that culminated in numerous deaths. In March 2006 Hamas began setting up its government in Gaza, which was followed by international sanctions. Public service delivery was severely weakened as a result of the internal fighting, international sanctions and the lack of administrative experience among Hamas cadre. Although Fatah and Hamas formed a unity government in March 2007, Hamas took control of Gaza in June 2007, taking over government offices and replacing Fatah officials. This was followed by a blockade of goods and people by Israel and the suspension of aid by most donors (the United States, the European Union, several European countries and the Arab states). Since the establishment of a Hamas government in Gaza, a transfer of projects took place from the 'old office' under the Palestinian Authority administration to the 'new office' under Hamas administration. The subsequent political tension and lack of communication between the old and new state institutions led to an extremely challenging context for case 4. Staff members from the Environmental Quality Authority and the PHG indicated that following a temporary suspension of their salaries after Hamas took government authority, payments to the PWA-staff were resumed under the condition that they would not resume their responsibilities, nor come to the office or take data and information with them. This transition was extremely challenging for NGOs and donors that were trying to continue their work.

The hostile and partly paralysed institutional context following from the institutional duality of the respective Fatah and Hamas administrative structures has affected, first and foremost, the *cognitions* of the MSP actors in the Bureij MSP that increasingly regarded the service provision project through a political (in fact conflictual) lens and the availability of resources. Communication and cooperation between both PWA and the Coastal Municipal Water Utility and the Al-Bureij municipality worsened from 2007 onwards. The Al-Bureij municipality council was replaced on 18 July 2007. Similarly, the community-based local water committee that was established in 2006 was very weak due to the internal conflict in June 2007. Therefore, a new local committee was established in 2007 that was selected from all blocks of the served area, and thus representing various views. According to one of the MSP members, nevertheless, re-establishment did not bring about smooth cooperation within the local water committee as 'the seeds of conflict were inside this committee'. Ultimately, this particular form of institutional multiplicity (or rather institutional polarization) also affected the *resources* available for the MSP due to withdrawal of international donors and underutilization of Fatah capacities.

The impact of the fragmentation in areas A, B and C

The West Bank is cantonized into various areas under different governance modalities. These areas followed from the Oslo Agreements and the areas were further demarcated by the Wye and Sharm el-Sheikh Memorandums. Area A (17 per cent of the land of the West Bank) is under Palestinian control; in Area B (24 per cent) civil affairs are under Palestinian control and security issues are under joint Palestinian–Israeli control; and Area C (64 per cent) is under Israeli civil and military control. Palestinians' access to water resources is limited by restrictions inherited from the Oslo Agreements. For example, they are not permitted to drill new wells in any of these areas or to extend existing wells without an Israeli permit, which is considered quite difficult to obtain (IDRC, 2008).

Cases 1 and 2 were particularly affected by the limited Palestinian jurisdiction and control over resources stemming from this administrative division. For both the Qalqiliya and Tulkarem irrigation well rehabilitation project and the Zahrat Al Finjan landfill in Jenin, several permits were required from the Joint Water Committee and the Israeli Civilian Administration. These permits were necessary for the choice of the location (case 2), for the design (case 2), for importing equipment (both) and for building new infrastructure (case 2). Yet, due to Jenin's vicinity to Israeli settlements, the Israeli authorities also recognized the importance of adequate waste management and the prevention of pollution of groundwater that the settlements also depend upon. As a result, the Israeli Environmental authorities were reported by MSP members to be quite helpful[10] and tried to advise the Palestinians on this project.

The Joint Water Committee has reportedly asked the Palestinian authorities, on several other occasions (not related to the four cases), to cooperate with the authorities administering the Israeli settlements on the West Bank for the establishment of new, or the rehabilitation of existing water and wastewater infrastructure – as a condition for permits. The Palestinian authorities tend to refuse such demands for obvious reasons: formal cooperation with the settlement authorities is perceived as legitimizing and recognizing settlements that are illegal according to international law.

The overall impact of fragility on Palestinian MSPs in the WASH+ sectors

Fragility in Palestine, as manifested in eruptions of violent conflict and institutional multiplicity, has had a range of negative effects on the governance and output of the MSPs studied. These effects were experienced both directly (in terms of damage to the infrastructure, the economy and mobility opportunities)[11] and indirectly (in terms of institutional arrangements which were the result of the Oslo Agreements and beyond). This was most apparent in case 4 in Gaza, which was never successfully implemented according to our respondents because of the political context described. Across the cases discussed, the negative effect was most apparent in the loss of resources through withdrawal of donors, underutilization of existing capacities, obstruction and delay in implementation processes and destruction of infrastructures and material. The wider fragile context also resulted in overly politicized (and often polarized) cognitions and motives regarding service provision in general and MSP cooperation in particular.

Concurrently, our interviews clearly showed that it was exactly because of the challenges and obstacles presented by conflict and institutional multiplicity that the inclusion of multiple stakeholders in WASH+ projects became a necessity and a partnership approach towards initiating and implementing such projects was regarded self-evidently desirable and was apparent in the majority of strategic motives of MSP actors. Our research also revealed that political unrest is not necessarily prohibiting or halting processes of cooperation between multiple stakeholders to improve service delivery. Case 2 for instance, the Zahrat Al Finjan project, was submitted only two weeks before the onset of the Second Intifada, and has – despite some delay – been completed and has reached far more beneficiaries (over 600,000) than originally intended (200,000). Moreover, case 1 in Qalqiliya and Tulkarem gained momentum (and eventual success) precisely because of the strategic relevance it acquired. The wider conflict, institutional multiplicity and lack of coherence characterizing the Palestinian WASH+ section, does not then have one singular effect, but plays out differently in different case-specific contexts. While fragmentation leads to more polarized motives, more uncertain cognitions and more zero-sum resource allocation in general terms (cf. Chapter 3), it can bring actors together and strengthen their ability to resolve as well.

The governance and implementation of the MSPs studied showed a remarkable flexibility, an ability 'not only to see the reality as a field of obstacles, but also as a terrain of potential – often unexpected – opportunities and being adaptive enough to use every "window of opportunity" to bring the ultimate purpose closer to realization' (Chapter 3 of this book). Processes of cooperation and social interaction between multiple stakeholders clearly changed during periods of armed conflict and, spurred by events such as the Second Intifada, actors applied a range of contingency and emergency planning measures with different degrees of success. The role of donors and their ground staff changed due to the tense relations between Israelis and Palestinians. A strong involvement of donors (in particular the European Union and the World Bank) during violent episodes in maintaining cordial dialogue with military authorities and pressure on the Joint Water Committee, was essential for the continuation of the activities and achievement of objectives of some MSPs.

9.6 Conclusion: the applicability of the MSP concept in fragile contexts

Synthesizing our above arguments, we arrive at three interrelated conclusions considering the applicability of MSPs as a tool for the governance of WASH+ projects in fragile settings.

First, partnerships including multiple stakeholders from the public and private sectors and civil society emerge in Palestine without explicit top-down policy guidance. This leads us to agree with Warner (cited in Nuijten, 2004, p. 117) that MSPs are perhaps more an empirical practice to be analysed than a policy prescription to be spread. The degree to which partnerships are consciously branded or instigated as MSPs do not seem to significantly affect the initiation, governance and implementation of MSPs (even if the MSP that was explicitly set up as an MSP did have the most extensive and sustainable output).

Second, MSPs do not emerge despite state fragility (in the wider context), but in many case exactly because of the multiplicity in actors, structures, regulations and practices (the structural context) that can be found in fragile situations. Respondents widely recognized the need to deal with flare-ups of violent conflict, competing Fatah and Hamas governments in Gaza and the administrative fragmentation of the West Bank respondents. This on-the-ground reality implies a need for:

1 national authorization of local projects;
2 Israeli permits and approval;
3 donor funding and assistance;
4 community contributions;

5 NGO lobbying and management;
6 the subsequent need for communication, exchange of information; and
7 joint management and decision making.

Third, although fragility might instigate the emergence of MSPs (for instance through adding a strategic aspect to practical motives), it simultaneously hampers their effectiveness and complicates governance and cooperation processes. This occurs as cognitions become polarized and politicized and resources diminish, remain unutilized or need to be used for conciliation rather than implementation. Thus, MSPs alone do not always guarantee a smooth process or a sustainable outcome. Problematic community relationships and conflicts can stall the process, in particular when stakeholders or beneficiaries have not been sufficiently involved in the process from the start.

The above conclusions have several implications for both the study and the support of MSPs in fragile contexts. Future focus need not so much be on promoting MSPs, as they emerge quite organically in one form or another. Donor organizations in the WASH+ sector, in Palestine at least, might benefit more from research into the potential for continuity during conflict and the dealing with hostile and contradicting authorities than from guidelines on initiation and internal stakeholder management. Moreover, instead of focusing on generalizing case-specific contexts towards a one-size-fits-all model, value might lie in exploring the diverse lessons learned on the effect of wider contexts on case-specific contexts and the different responses of stakeholders to such wider political developments.

We could not agree more with Bressers and de Boer, who state in Chapter 3 that 'understanding the various impacts from various perspectives and influencing factors – particularly in the complex policy realm of water governance – is highly dependent on context', and we hope to have convincingly argued that in 'knowing "what works, where, when and how in transferring lessons in water management"' more attention needs to devoted to the fragile political contexts in which many WASH+ projects are set.

Notes

1 See O'Riordan and Stoll-Kleeman (2002) on participatory planning and deliberative democracy; Ostrom (1990) and Mayntz (1993) on self-regulating institutions and Uphoff (1996) on participatory management (Nuijten, 2004, p. 116).
2 Findings and conclusions are discussed in detail in a report (Van der Molen *et al.*, 2011) available on the PSDN website: www.psdnetwork.nl.
3 The Green Line was agreed upon as part of the Armistice Agreements of 1949, between Israel and the countries around Israel (Egypt, Lebanon, Jordan and Syria).

4 The project reached 30,044 beneficiaries, 243,000 m^2 of land was reclaimed in area C and the Seam Zone; 19,505 additional m^3 water available for domestic use/month; 37 per cent reduction in operational costs; 99.6 per cent reduction of water supply interruptions, increase of average abstraction capacity from 57.5 m^3/hour to 104.8 m^3/hour; pumping capacity increased by 82 per cent and 93 per cent (Van der Molen *et al.*, 2011, p. 41).

5 It goes beyond the purpose of this study to include a more elaborate discussion of the academic debate on fragility. A comprehensive overview can be found in Van Overbeek *et al.* (2009).

6 See: http://siteresources.worldbank.org/INTWESTBANKGAZA/Resources/Annex12,Aug31,2009(enriched).pdf; http://siteresources.worldbank.org/INTWESTBANKGAZA/Resources/WaterRestrictionsReport18Apr2009.pdf; and http://go.worldbank.org/71X59QSH80, accessed 1 December 2011.

7 The high price of water contributed to the abandonment of land by 123 farmers, resulting in 570,000 m^2 (570 dunums) uncultivated, of which 480 dunums in area C, and ninety dunums in the seam zone (ECHO, 2010, p. 8).

8 The project, managed by the MSP, was effective in achieving both objectives, by providing an additional 19,505 m^3/month for domestic water supply, by 37 per cent reduction of operational costs, by 99.6 per cent reduction of water supply interruptions, and by reclaiming 243 dunums of land in area C and in the seam zone. It was less successful in guaranteeing sustainability.

9 The damage to the solid waste infrastructure in Jenin and Tubas governorates resulting from actions by the Israeli Defence Forces in the West Bank between March and May 2002 was estimated at US$169,820. Operation Cast Lead of December 2008/January 2009 resulted in damage of water and wastewater infrastructure in the Gaza Strip, with an estimated cost of US$6 million (CMWU, 2009).

10 Cooperation may also have been facilitated by the Joint Services Council decision to hire an Israeli solid waste management expert for redesign of the landfill and supervision of the construction.

11 Akkaya *et al.* (2011) elaborately discuss the detrimental effects of the closures regime (consisting of continuous, but unannounced physical and administrative road blocks, checkpoints, barriers) on the Palestinian economy. Following Diwan and Shaban (1999) and Fischer *et al.*(2001) and in line with UNESCO data, Akkaya *et al.* estimate that each day of closure costs the Palestinian economy US$7 million in terms of lost income.

References

Akkaya, S., Fiess, N., Kaminski, B. and G. Raballand, G. (2011) 'Fragility and conflict in Palestine: The costs of the closures regime on West Bank and Gaza', in W. Naudé, A.U. Santos-Paulino and M. McGillivray (eds) *Fragile States: Causes, Costs and Responses,* Oxford University Press, Oxford.

Boege, V., Brown, M. and Clements, K. (2009) 'Hybrid political orders, not fragile states', *Peace Review,* vol. 21, no. 1, pp. 13–21.

CMWU (Coastal Municipalities Water Utility) (2009) *Damage Assessment Report 'Water and Waste Water Infrastructure and Facilities' Gaza, 27 Dec. 2008–19 Jan. 2009,* online, available at: www.ochaopt.org/documents/opt_wash_cmwu_Water_Wastewater_Infra_Damage_jan_2009.pdf, accessed 29 October 2012.

De Boer, D. and Pfisterer, S. (2009) 'The missing link in state-building: Bilateral donor and donor NGO policies on linking state-society building in fragile states', in G. Molier and E. Nieuwenhuys (eds) *Peace, Security and Development in the Era of*

Globalization: The Integrated Security Approach viewed from a Multidisciplinary Perspective, Leiden University, Leiden.

DfID (Department for International Development) (2011) *State-Building, Peace-Building and Service Delivery in Fragile and Conflict Affected States: Literature Review: Final Report*, DfID, London.

Diwan, I. and Shaban, R.A. (eds) (1999) *Development Under Adversity: The Palestinian Economy in Transition*, 489, Palestine Economic Policy Research Institute (MAS) and the World Bank, Washington, DC.

ECHO (European Commission Directorate General for Humanitarian Aid) (2010) *Single Form for Humanitarian Aid Actions: Final Report – version 14/06/2010*, ECHO, Brussels.

EHP (Environmental Health Project) (2002) *Village Waste Water System Feasibility Report for the West Hebron and South Nablus Clusters: Final Report*, prepared for the USAID Mission for the West Bank and Gaza under Task Order HRN-1-802-99-00011-00.

Fischer, S., Alonso-Gamo, P. and Erickson von Allmen, U. (2001) 'Economic Development in the West Bank and Gaza since Oslo', *The Economic Journal*, vol. 111, no. 472, pp.254–275.

Grimsey, D. and Lewis, M. (2004) 'The governance of contractual relationships in public–private partnerships', *The Journal of Corporate Citizenship*, vol. 15, pp. 91–109

Hemmati, M. (2002) *Multi-Stakeholder Processes for Governance and Sustainability. Beyond Deadlock and Conflict*, Earthscan, London.

Hofmann, B. (2009) 'Are hybrid political orders an appropriate concept for state formation? Timor-Leste revisited', in *Berghof Handbook Dialogue*, No. 8, Berghof Research Center for Constructive Conflict Management, Berlin.

IDRC (International Development Research Centre) (2008) 'Water in Palestine', formerly available at: www.idrc.ca/en/ev-65880-201-1-DO_TOPIC.html, accessed 3 October 2008.

Mayntz, R. (1993) 'Governing failures and the problem of governability: Some comments on a theoretical paradigm', in J. Kooiman (ed.) *Modern Governance: New Government–Society Interactions*, Sage, London.

Multipart (2008) 'Multi-stakeholder partnership in post-conflict reconstruction: the role of EU, Theoretical & methodological framework and guidance for the project', online, available at: www.humansecuritygateway.com/documents/MULTIPART_MultiStakeholderPartnerships_PostConflictReconstruction_RoleOfEU.pdf, accessed 11 February 2009.

Naudé, W., Santos-Paulino, A.U. and McGillivray, M. (eds) (2011) *Fragile States. Causes, Costs and Responses*, Oxford University Press, Oxford.

Noor, M. (2010) 'Multi-stakeholder processes', in M. Noor, N. Douma, G. van der Haar, D. Hilhorst, I. van der Molen and N. Stel *Multi-Stakeholder Processes, Service Delivery and State Institutions. Theoretical Framework and Methodologies*, Working Paper, Network for Peace, Security and Development, the Netherlands, pp.34–49, online, available at: www.psdnetwork.nl/documenten/publications/20100301_Theoretical_Framework_&_Methodologies_.pdf.

Noor, M., Douma, N., Van der Haar, G., Hilhorst, D., van der Molen, I. and Stel, N. (2010) *Multi-Stakeholder Processes, Service Delivery and State Institutions. Theoretical Framework and Methodologies*, Working Working Paper, Network for Peace, Security and Development, the Netherlands, online, available at: www.psdnetwork.nl/

documenten/publications/20100301_Theoretical_Framework_&_Methodologies _.pdf.

Nuijten, M. (2004) 'Governance in action: Some theoretical and practical reflections on a key concept', in D. Kalb, W. Pansters and H. Siebers (eds) *Globalization and Development: Themes and Concepts in Current Research,* Kluwer Academic Publishers, New York, pp. 103–130.

OECD (Organisation for Economic Co-operation and Development) (2008) 'Service Delivery in Fragile Situations. Key concepts, findings and lessons', *Journal of Development,* vol. 9, no. 3, pp. 3–54.

OECD (Organisation for Economic Co-operation and Development) (2012) *Meeting the Water Reform Challenge,* OECD Studies on Water, OECD Publishing, Paris.

O'Riordan, T. and Stoll-Kleeman, S. (eds) (2002) *Biodiversity, Sustainability and Human Communities: Protecting Beyond the protected,* Cambridge University Press, Cambridge.

Ostrom, F. (1990) *Governing the Commons: The Evolution of Institutions for Collective Action,* Cambridge University Press, Cambridge.

Pierre, J. and Peters, G.B. (2000) *Governance, Politics and the State,* St. Martin's Press, New York.

Plummer, J. and Slaymaker, T. (2007) *Rethinking Governance in Water Services,* Working Paper 284, Overseas Development Institute, London.

Sørensen, E. and Torfing, J. (2005) 'The democratic anchoring of governance networks', *Scandinavian Political Studies,* vol. 28, no. 3, pp.195–218.

Stel, N.M. (2012) 'Conference report service provision and the legitimacy of state institutions in situations of conflict and fragility – The Hague, 7 December 2011', Maastricht School of Management, Maastricht, online, available at: www.psdnetwork.nl/documenten/publications/20120103_Conference_Report_Service_Provision_and_the_Legitimacy_of_State_Institutions_in_Situations_of_Conflict_and_Fragility.pdf.

Stel, N.M., De Boer, D. and Hilhorst, D. (2012) 'Multi-stakeholder processes, service provision and the legitimacy of state institutions in situations of conflict and fragility: Experiences from Burundi, DR Congo, Nepal and the Palestinian Territories', Synthesis Report of the working group 'MSPs, Service Delivery and State Institutions' of the Peace, Security and Development Network, online, available at: www.psdnetwork.nl/documenten/publications/20120202_Synthesis_Report_Multi_Stakeholder_Processes_Service_Delivery_and_State_Insitutions.pdf.

Tropp, H. (2005) 'Developing water governance capacities', feature article, UNDP Water Governance Facility/SIWI, Stockholm.

United Nations Development Programme (2004) *Water Governance for Poverty Reduction: Key Issues and the UNDP Response to the Millennium Development Goals,* UNDP, New York.

Uphoff, N. (1996) *Learning from Gal Oya, Possibilities for Participatory Development and Post-Newtonian Social Science,* Intermedia Technology Publications, London.

Van der Molen, I., Hasan, W. and Tamimi, A. (2011) *Multi-stakeholder Processes, Service Delivery and State Institutions: Experiences from the West Bank and Gaza Strip, Palestinian Territories,* Report of the Working Group on Multi-Stakeholder Processes, Service Delivery and State Institutions of the Peace, Security and Development Network in cooperation with the Palestinian Hydrology Group.

Van Overbeek, F., Hollander, T., Van der Molen, I., Willems, R., Frerks, G. and Anten, L. (2009) *The Fragile States Discourse Unveiled,* Working Paper 1, Peace, Security and Development Network, Utrecht.

van Tongeren, P. and van Empel, C. (eds) (2007) *Joint Action for Prevention Civil Society and Government Cooperation on Conflict Prevention and Peacebuilding,* Issue Paper 4, European Centre for Conflict Prevention/Global Secretariat of the Global Partnership for the Prevention of Armed Conflict.

Vermeulen, S., Woodhill, J., Proctor, F. and Delnoy, R. (2008) 'Chain-wide learning for inclusive agrifood market development: A guide to multi-stakeholder processes for linking small-scale producers to modern markets', International Institute for Environment and Development (IIED), UK and the Capacity Development and Institutional Change Program (CD&IC), Wageningen University and Research Centre, the Netherlands.

Warner, J.F. (2006) 'More sustainable participation? Multi-stakeholder platforms for integrated catchment management', *International Journal of Water Resources Development,* vol. 22, no. 1, pp. 15–35.

World Bank (2009) *West Bank and Gaza: Assessment of restrictions on Palestinian Water Sector Development, Sector Note,* Report No. 47657-GZ, World Bank, Middle-East and North-Africa Region.

10 Transferring international commitments to the local level

The case of integrated urban wastewater management in Hanoi, Vietnam

Jaap Evers

10.1 Introduction

International commitments on green sustainable growth often encourage the movement towards recycling societies in which valuable components of waste products are recovered and reused. Wastewater is one such product that is used for irrigation, both in arid and humid regions. It is estimated from a survey across the developing world that wastewater without any significant treatment is used for irrigation in four out of five cities (Raschid-Sally and Jayakody, 2008). Wastewater use in (peri)urban agriculture transcends the typical sectoral and geographical policy boundaries of urban water management. Planning for wastewater use requires the involvement of many agencies covering health, water, sanitation, irrigation and agriculture as well as the users, knowledge institutes and private sector. Urban and peri-urban agriculture is frequently tied to sustainable urban development (Smit and Nasr, 1992; Drechsel *et al.*, 2010). With all its limitations, the potential of (peri)urban agriculture to reuse urban wastewater (to recover valuable nutrients) and to be part of urban sanitation systems as environmental service providers is substantial. Yet, many policymakers at the local level are unaware of the significance of this practice and the possible opportunities it can bring (Drechsel *et al.*, 2010).

Agenda 21, the key output of the 1992 UN conference on Environment and Development, recognizes the vital role of local authorities in realizing sustainable use of natural resources (United Nations, 1992). This and many other international commitments underline the importance of the role that urban agriculture can play in urban policy areas like economic development, health, food security, urban environmental management and social development. As globally most urban wastewater is discharged untreated, the practice of wastewater irrigated agriculture is widespread and includes the use of polluted surface water. Although the use of wastewater in urban agriculture is common, this is mostly informal and uncontrolled, and regularly has negative effects on people's health and water resources (Drechsel *et al.*, 2010; Raschid-Sally and Jayakody, 2008; Evers *et al.*, 2010).

Integrated Water Resources Management (IWRM) is proposed as an approach which can support overcoming the problems inherent to uncoordinated, sector-dominated and competitive water management (Rees, 2006). Transferring (and converting) philosophies on green economies and IWRM into practice is a challenge – which is in many cases understating the real difficulties seen in implementation. As mentioned earlier in Chapter 2 and by Biswas (2004) the use of IWRM in practice has been minimal or even indiscernible. Due to the large gap between theory and practice, IWRM should be regarded as a management philosophy that 'promotes the coordinated development of and management of water, land and related resources, in order to maximize the resultant economic and social welfare in an equitable manner without compromising the sustainability of vital ecosystems' (GWP, 2000). Institutional integration of nutrient recovery from wastewater sources in urban systems is therefore a strong indicator of effective implementation of elements of the IWRM philosophy and local political will towards realizing green urban economies.

This chapter addresses the question: why are international concepts (like the Ecosan concept discussed by Werner *et al.*, 2004) and commitments (like in the Millennium Development Goals) on sustainable cities hardly transferred into integrated urban water resources management policies and practices in metropolizing cities in the developing world? 'Sustainable cities' here refers to cities which are managed with the focus on minimizing their current and future environmental impact. The practice of recovering the valuable resources from wastewater by using it for agricultural production is recognized as key in realizing green sustainable urban environments. When wastewater use occurs through a formal practice of wastewater irrigation it reveals functionally integrated urban water management systems. To address this issue, this chapter examines the case of Hanoi, Vietnam. Hanoi has adopted a long-term strategy to turn itself into a 'green, civilized and modern city' by 2050, which will involve setting aside up to 70 per cent of the city's natural territory for 'tree and water space' (EIU, 2011). As most urbanization is predicted to take place in developing countries, Hanoi is regarded as a typical metropolizing city in a country in economic transition, with all its governance challenges on managing population growth, food security, housing, infrastructure, socio-economic welfare, health, sanitation, environmental health and urban (waste) water systems. National and local governments are limited by their resources (like power and finances). Based on perceptions and political motivations on urban development, governments – administrators, politicians, and civil servants – therefore allocate resources to their priorities in urban development. At the same time their influence on (sustainable) urban development is limited and influenced by all sorts of contextual developments. This is described by the actor characteristic and the contextual layers in social interaction processes by the Contextual Interaction Theory, as elaborated in Chapter 3.

This chapter finds its origin in a research project of the author in Hanoi (Evers, 2006). During field research, semi-structured open interviews were held at three institutional levels (communal, district and municipal) to assess the level of integration of peri-urban irrigated agriculture in the urban wastewater management system and its effects on sustainable urban development. The governmental stakeholders at district and municipal level were defined based on an institutional system analysis by Luan and Minh (2005). At the communal level interviews and observations were carried out at Đong Du commune in the peri-urban district Gia Lam. Here representatives of the communal people's committee, farmers' union, and farmers' cooperative were interviewed. In addition semi-structured open interviews were conducted with farmers, which were triangulated by a group discussion. This commune was representative for the issue as farmers in this commune lost land to the Thanh Tri bridge construction project and urbanization processes in Gia Lam district also led to deterioration of irrigation water quality. Water samples from the irrigation system showed that the nutrient concentrations in the irrigation water were high enough to at least partly fulfil crop nutrient requirements (depending on the crop). This chapter uses this research and follow-up work by Van der Water (2011) in the same commune as example to analyse the question posed earlier by using elements of the Contextual Interaction Theory.

The following section introduces the importance of cities in sustainable development, the variety of international commitments on this issue, and the importance of IWRM to realize the goals of these commitments. This will be followed by a discussion on the phenomenon and potential of urban wastewater irrigation within urban water governance systems. The question will then be further analysed via the example of Hanoi. In the last section conclusions from the case are drawn and recommendations are made for effective transfer.

10.2 Cities, sustainable development and international commitments

This section highlights the challenges of an urbanizing developing world, the accompanying increase of wastewater effluents, and the international commitments that exist to address these issues. IWRM is explored as a process for realizing sustainable urban environments. The international level is chosen as the context where the relevant concepts on sustainable urban systems and IWRM are developed and promoted.

Urban populations are projected to increase from the current 3.4 billion to over 6.4 billion by 2050, which indicates that two-thirds of the global population will be urbanized (Corcoran *et al.*, 2010). Nearly all of this growth will be in developing countries (UNPF, 2011). Currently, 90 per cent of the wastewater effluents are discharged untreated (Corcoran *et al.*, 2010) and in many cities in the developing world these effluents are used directly and indirectly (diluted with other water sources) in informal urban agricultural systems (Raschid-Sally and Jayakody, 2008). These interactions highlight why cities have a central role to play in the sustainable development of our world. As centres of social interaction and economic activity, cities are critical for the formulation and implementation of policies across sectors for sustainable development (UN-Water, 2011).

Unprecedented urbanization rates outpace the capacity and ability of urban planners to transfer and implement international ideas on sustainable growth and green cities, despite the billions of dollars that are being invested (UNPF, 2011). This is best represented by the many informal housing areas (slums) in megacities in the developing world, where basic services of water and sanitation are often missing. Since the Rio Earth Summit in 1992, many positive developments have been experienced in many parts of the world. However, due to intense urbanization the provision of basic water and sanitation, and associated institutions (as addressed in Agenda 21), are still of great concern today (UNCSD, 2011). The lack of proper sanitation results in environmental degradation and is directly related to diarrhoeal diseases and child mortality (Corcoran *et al.*, 2010). The international efforts have seen great successes in air quality issues, transport and access to clean drinking water. Serious challenges remain however in terms of access to proper sanitation and sewage handling, especially in informal urban settlements (UNCSD, 2011).

Urban agriculture can make an important contribution to the sustainable development of cities, in both the developed and the developing world (Drechsel *et al.*, 2010). Its potential for realizing sustainable cities was already recognized twenty years ago, for example by Smit and Nasr (1992). Despite increasing attention and a growing number of cities interested in action programmes on urban agriculture for reuse and recovery of valuable nutrients, wastewater use in urban agriculture still remains a relatively new field of research. Debates continue on its viability, relationship to pollution, and contribution to urban economic development, as well as the need for political support (FAO, 2007).

After the Rio Earth Summit, the importance of cities for sustainable development has been addressed in a number of international commitments. The urban dimension is also present in the Millennium Development Goals (see Table 10.1). Cities have been featured in international commitments covering many topics in addition to water, sanitation and sewage, such as slums, transport, health, access to clean and safe energy, and local government capacity. Table 10.1 gives an overview of selected commitments, their targets and delivery dates. Although many positive developments in realizing sustainable cities have been registered since 1992, most of the issues are still of great concern due to unprecedented urbanization rates.

Table 10.1 Overview of international commitments related to cities with time-bound targets

Commitment	*Issues or targets addressed*
Agenda 21	Protecting and Promoting Human Health. Meeting the urban health challenge. Promoting Sustainable Human Settlement Development. Promoting the integrated provision of environmental infrastructure: water, sanitation, drainage and solid-waste management. Promoting human resource development and capacity building for human settlements development. Environmentally sound management of Solid Wastes and Sewage related issues Local Authorities Initiatives in Support of Agenda 21.
Millennium Development Goals (MDG)	Goal 7 – Ensure Environmental Sustainability: Halve, by 2015, the proportion of the population without sustainable access to safe drinking water and basic sanitation. By 2020, to have achieved a significant improvement in the lives of at least 100 million slum dwellers.
Johannesburg Plan of Implementation	Chapter II – Poverty Eradication: To halve, by the year 2015, the proportion of people who are unable to reach or to afford safe drinking water and the proportion of people who do not have access to basic sanitation. By 2020, achieve a significant improvement in the lives of at least 100 million slum dwellers, as proposed in the 'Cities without slums' initiative.
Plan of Action on Subnational Governments, Cities and Other Local Authorities for Biodiversity (2011–2020)	By 2020 governments need to have installed: • Relevant tools, guidelines and capacity building • National biodiversity strategies and action plans, • Monitoring and evaluation systems for subnational governments and local authorities, guided by national frameworks, to report on progress to national governments in line with reporting obligations under the Convention on Biological Diversity, and to set benchmarks for local biodiversity management in line with the Convention on Biological Diversity, using tools such as the Singapore Index on Cities' Biodiversity.

Source: UNCSD (2011).

Sectoral approaches lead to suboptimal outcomes at the water system level (Barua and Van Ast, 2011). This also holds true for the water basin and the urban system. Fragmented and sectoral urban planning leads to suboptimal outcomes. Within the IWRM perspective, four broad reasons justify why IWRM processes are essential for sustainable urban growth (Rees, 2006, p. 12–13):

1. The negative externalities that arise from the uncoordinated use of the physically interdependent water and land resources.
2. The opportunity costs which arise when factors of production (including water, land and capital) are employed for low value/benefit purposes.
3. The negative externalities and opportunity costs which arise from the uncoordinated provision (non-provision) of interdependent basic services such as health, education and sanitation.
4. The cost savings that can occur by widening the range of provision or management options.

The above reasons make sense from a theoretical perspective. However in practice, as was referred to earlier from Biswas (2004), true IWRM processes (even when labelled as such) are rare, because urban water governance systems have diverse stakeholders (industries, housing corporations, urban slum dwellers, peri-urban farmers) with uneven power balances and legal boundaries for the management of surface water and groundwater; water supply, drainage, wastewater and agricultural water. An additional hurdle to the implementation of IWRM processes is the lack of integration between 'water institutions' and institutions representing other sectors, like transport, infrastructure, housing, or public health into IWRM processes. When implementing agencies cling rigidly to the principles of IWRM, processes are likely to come to a standstill (Geldof, 2001; Evers, 2011). In the end, power relations are not easily balanced and legal institutions are not easily changed. Implementation of integrated urban (waste) water management plans requires much bricolage, which means to transfer or develop new governance systems using available elements to create context specific governance structures (Sehring, 2009). At the same time it requires flexibility, adaptive capacity, brokerage and therefore 'learning while doing' approaches from the governance network actors to be effective (Evers, 2011). It is thus important to have a shared vision on sustainable urban development among the different stakeholders, who are acting in various parallel urban development processes, so that decentralized processes do contribute to sustainable development without trying to control it as being one process (Evers, 2011).

10.3 Wastewater use in urban agriculture and urban water management: the extent of the issue

Although wastewater irrigation is recognized as a common practice, there is no comprehensive global inventory of the extent to which treated and non-treated wastewaters are used for irrigation. There are several international commitments and concepts on sustainable urban development and developing green cities. Sustainable urban development requires these international commitments to be transferred to the local contexts of cities. This section discusses elements that explain the successfulness of this transfer.

Jiménez *et al.* (2010) estimate that globally at least 4 to 6 million hectares of land are irrigated with either wastewater or polluted water. The WHO (2006) estimates it to be as much as 20 million hectares, almost 8 per cent of the global irrigated area (Bruinsma, 2011). The area irrigated with treated wastewater amounts to only a tenth of this value (Jiménez *et al.*, 2010). Because wastewater is often used in informal settings estimations are difficult to make and the usage can actually be much higher. Much of this use is not intentional, but a consequence of water sources being polluted due to poor sanitation and wastewater disposal practices in cities. Based on a survey across the developing world, Raschid-Sally and Jayakody (2008) suggest that wastewater is used for irrigation purposes in four out of five cities. In the cities of developing countries, the main drivers for wastewater irrigated (peri)urban agriculture are a combination of: (1) limited capacities of cities to treat their wastewater; (2) a lack of alternative water sources; and (3) urban food demand and market incentives favouring food production in the proximity of cities (Raschid-Sally and Jayakody, 2008).

Although the use of wastewater in agriculture is accompanied with human and environmental health risks due to the chemical compounds from industries and pathogens from urban wastewater, it also has valuable advantages. The first advantage of wastewater reuse is its nutrient content. Even when treated, wastewater recycles organic matter and a larger diversity of nutrients than any commercial fertilizer can provide. Evers (2006) shows that farmers in Hanoi could reduce their application of chemical fertilizers significantly by using these nutrient-rich sources of water for irrigation. Global fertilizer production is responsible for about 1 per cent of the world's total energy use (which has grown at an average rate of 3.8 per cent per year) and although significant energy efficiency improvements have been realized in the production of fertilizers (except for phosphate fertilizers) these improvements have not been sufficient to offset growing energy demand due to rising fertilizer consumption (Ramírez and Worrel, 2005). With a continuing increase in energy prices (and fertilizers), wastewater offers cheap nutrients and therefore increases the economic benefits to farmers. As phosphorus is a depleting finite resource, excreta and wastewater can be critical sources of phosphorus. By using nutrients from wastewater effluents, agriculture has the potential to play an important part in urban wastewater management and the environmentally safe discharge of effluents into the environment. Another important factor that makes wastewater valuable is that it is a reliable source of water, as it is available all year round, unlike precipitation or seasonal streams.

Urbanization is often accompanied with changing (higher value) diets that include fewer grains, yet more meat and vegetables. Since vegetables are highly perishable, this is an incentive to grow them in the vicinity of cities. It is estimated that (peri)urban agriculture in Hanoi provides 60–80 per cent of the perishable vegetable products for the local market (Van den Berg *et al.*, 2003) and that freshness is the advantage of peri-urban vegetables sold in the Hanoi markets (Figuié, 2004). When vegetables are the main commodity produced with wastewater, there can be a significant aggregate benefit for the society in terms of a more balanced diet to urban citizens and to farmers through higher incomes (vegetables are known as cash crops as opposed to staple crops such as cereals). This change in market demand and supply by urban farmers occurs as rapidly as urbanization itself (and in the Vietnamese case is combined with rapid economic growth). Where Evers (2006) describes a fifty–fifty mixed cropping system of rice and cash crops (herbs, tomatoes and some fruit trees) in Đông Dư commune in Hanoi, Van der Water (2011) observes five years later a system of mainly cash crops (mostly fruit trees and herbs) with some patches left for rice production. Van der Water also observes a shift to using groundwater for irrigation. The main source of irrigation, Cầu Bây River, which receives about 13,000 m^3 domestic and industrial wastewater per day, is regarded as too polluted by most farmers. Although excessive chemical fertilizer applications in 2006 decreased to proper levels in 2011, this still means little recovery of nutrients from the wastewater. Land application of wastewater for agricultural use constitutes a low-cost disposal method and land-treatment system. When carried out under controlled conditions, it can be safely implemented and recover nutrients for agricultural production (Jiménez *et al.*, 2010).

Globally irrigated agriculture is the largest user of freshwater (Molden, 2007). However when water managers need to choose, water is mostly allocated to the highest value use (domestic, industrial, agricultural and then environmental). In the context of urbanization, the competition over water resources among different water users will increase. This might not be as dramatic as it sounds, as most of the urban water users discharge the majority of the water used back into the system as wastewater. The wastewater effluent becomes available to other users, like agriculture, which does not demand as high a quality of water as urban water use. While agriculture at a basin level is always counted as a water user, this is not reflected in urban water management. Water can be used and allocated multiple times by different users from the highest to lowest quality-demanding user, which mostly corresponds with the highest to lowest economic valued use. Integrated urban water management in practice is often considered more or less similar to an integrated approach to secure drinking water supply, sewer management and flood prevention. It often uses demand management instruments for urban water supply and sanitation services, which highlights potential tensions between urban water uses, but neglects the interrelationships between water and other services (Rees, 2006). Within an urbanizing world and the demand for realizing green cities and food security for urban populations this needs to be linked up with dimensions of sustainable agriculture: increased productivity, reduced risk, increased resource protection, economic viability and social acceptability (Penning de Vries *et al.*, 2002).

Even though the potential value of wastewater effluents for agriculture is well known (see also the Ecosan concept of Werner *et al.*, 2004) in most regions, there is very little planned and formal use and safe recovery (for both people and environment) of wastewater (Drechsel *et al.*, 2010). Additionally, large centralized sewerage systems are very expensive to operate and maintain and thus many systems in the developing world are prone to mismanagement and malfunctioning. Low-tech, decentralized systems are therefore preferred (Huibers *et al.*, 2010). The issue of safe wastewater use in urban irrigation for realizing more sustainable cities is not a question of engineering impossibilities, but is rather related to social and governance issues. While the growth of complexity (the extent of issues incorporated) in integrated water management regimes seems to be part of a more general development, more coherence among these elements is not (Bressers and Kuks, 2002). Therefore, integration as it is used in the policy sphere is a combination of extent and coherence. Evers *et al.* (2010) compare this same coherence concept with supply chain management theories in which they refer to the development towards more coherence as moving towards vertical integration and describe the development towards fragmentation as spot market behaviour. The latter means that stakeholders solely focus on their personal short-term benefits. Based on the governance regime structure described by Bressers and Kuks (2003, 2005) and the case of Hanoi we will try to get a grasp on the complexity of transferring these international commitments on green cities and urban IWRM to the local level.

10.4 Wastewater irrigation and integrated urban water resources management in Hanoi

Hanoi is the capital and the second largest city of Vietnam and currently has a population of 6.5 million. About 2.7 million people live in the urban districts, while the remaining 3.8 million live in the peri-urban or rural districts (GSO, 2009). The urban sewerage system was built between 1905 and 1945 (during the French colonial period). Only an estimated 40 per cent of Hanoi's residents have access to a sewerage connection and about 10 per cent of the wastewater effluent is treated (EIU, 2011). Hanoi's centralized sewerage and drainage system is outdated and its capacity is insufficient for the city's current population. Hanoi does not (yet) have a plan to promote environmentally sustainable sanitation services. The authorities are tightening regulations regarding the treatment of industrial wastewater, and they are increasingly fining offenders (EIU, 2011). In mid-2010, the Hanoi People's Committee announced that new industrial parks in the city would be required to have wastewater treatment facilities and existing parks must treat their own waste as well. Hanoi's authorities have also embarked on a major campaign to clean the city's heavily polluted rivers and lakes. In early 2010, the authorities announced that they would spend US$81 million, backed by domestic private firms, to clean up forty-five lakes by 2015, and work has already begun on some of the city's largest lakes (EIU, 2011). A number of initiatives are being taken to halt the deterioration of the urban and surrounding environments of Hanoi. In 2011, the government approved a master construction plan for the development of Hanoi towards 2030, although it is actually more of a vision than a construction plan. It includes a green belt of parks, lakes and land for agriculture between the old city core and five satellite cities for urban expansion. The plan demands an enormous overhaul of Hanoi's water supply, sewerage, transport etc. infrastructure and will affect many people's lives. Although this vision is backed by the national government, the implementation of any future green initiative can still be undermined by the national focus on economic growth and uncontrolled urban development.

Hanoi is directly subject to state authority and therefore has a four-level governmental structure. Every level (state, municipal, district and communal) has its own people's committee and most of the ministries have line agencies at the municipal and district levels. Representative departments at lower levels have to refer their problems to higher levels. The higher-level organizations make decisions for implementation at lower levels in a very time consuming process. Luan and Minh (2005) note in their work on agricultural institutional aspects in Hanoi, that the system lacks a synchronized coordination between, on the one hand, agencies that make decisions (higher-level departments) and on the other, agencies that are responsible for implementation (lower-level departments). Responsibilities concerning the various elements of integrating agriculture in urban water management are divided across different departments. In addition, there is a spatial separation of departments in the urban and the peri-urban districts of Hanoi. Although one of the basic ideas of IWRM is that water decisions should be made at the lowest level, in a fragmented (and/or hierarchical) governance system this can increase difficulties in coherent decision-making.

Table 10.2 gives an overview of related stakeholders within the hierarchical levels of the municipal boundaries of Hanoi (Evers, 2006). These stakeholders have their own instruments to influence urban water governance based on their perceptions on wastewater, urban water management, food production, health etc. All of these actors influence the level of environmental sustainability of the urban system.

In line with the general coherent governance structure outlined in Chapter 3, this is envisioned in this case as having all actors in the urban wastewater system share the responsibility for the environmentally sound discharge of waste effluents and coordination of their activities to realize a common goal. Policy goals and instruments are complementary and support implementation effectively to address the environmental, economic and social issues. To actively cooperate with other stakeholders in reaching a common goal, they need to first each recognize their dependence on each

Table 10.2 Overview of stakeholders in integrated urban wastewater management in Hanoi

Municipal Level
Hanoi's People's Committee
Department of Science and Technology
Department of Natural Resources and Environment and Land
Department of Agriculture and Rural Development
Department of Public Works
Urban Environmental Company
Department of Urban Planning
Department of Health
Hanoi Drainage and Water Management Company
Department of Economic Planning and Rural Development
Farmers Union
Hanoi Water Works Company
District level (eleven urban districts, eighteen peri-urban/rural districts)
Districts People's Committee
Agriculture Extension Centre
Irrigation Company
District offices of municipal departments
Communal level (>500 wards)
Communal People's Committee
Communal Farmers Union
Farmers Cooperative
Health service
Committee on Flood Prevention and Sanitation
Farmers
Household as wastewater users, food producers and possible polluters
General
Urban residents as water users and producers of wastewater
Urban residents as consumers of locally produced food
Industries

other. A high level of mutual recognition between stakeholders means that there is knowledge of how the system is managed, even when actors are not cooperating, communicating or coordinating activities. However, as also described by Molle and Hoanh (2009), the lack of institutional clarity often induces competition among departments over tasks and responsibilities, and at the same time causes gaps.

A high level of fragmentation exists in the governance system of urban wastewater management and peri-urban agriculture in Hanoi. Institutionally, urban wastewater management and agriculture are strictly separated via the responsible government bodies (see Table 10.2) and officials often deny the presence of wastewater irrigation in urban agriculture. This is understandable from their perspective, as formally surface water is used for irrigation and its pollution is regarded as a problem. That the irrigation water is in fact diluted or sometimes raw sewage (undiluted and non-treated) might be beyond their scope and might also be a matter of perception. With regard to wastewater management, responsibilities are scattered among different departments. No department is fully responsible for (peri)urban wastewater management, at the same time officials do not fully comprehend the responsibilities of other departments (Raschid-Sally *et al.*, 2004; Evers *et al.*, 2010). Officials lack knowledge about the responsibilities of other governmental bodies and do not regard farmers as participants in the planning and implementation process. The latter might not be problematic when farmers are well represented by their local people's committee or the farmers union. However, in the centrally organized hierarchical system, there is little room for civil society to participate in policy development. As participation of stakeholders is essential to IWRM and sustainable development, this level of fragmentation of the governance system is particularly alarming since the extent of the issue is growing.

The results of this fragmented system are also visible at the field level. The relations between farmers and the administrative structure are weak. This is exemplified by the low level of knowledge that farmers have regarding the institutional structure, as well as their on-the-ground agricultural practices. One of the major problems at field level is the general lack of knowledge of farmers on proper nutrient management. Farmers generally stated that neighbouring farmers and friends are the most important sources of information. One would think then that all farmers in the commune would use more or less the same amounts of fertilizer. In reality, however, among farmers in Đông Dư commune nitrogen fertilizer applications differed by tenfold (Evers, 2006). Apparently, farmers do not have much trust in their information sources to take the risk of yield reduction by reducing fertilizer inputs. As a consequence, the combination of nutrient-rich irrigation water and often excessive application of fertilizers makes agriculture an extra source of pollution instead of a step in environmental sanitation. Finally, farmers do not recognize themselves as co-managers of the water system as they are not aware of the negative consequences of their habits for downstream activities, nor are they implementing improved nutrient management practices. Farmers act in self-interest, while sharing only little information with other stakeholders; further damaging is that the information shared is minimally trusted. It is not surprising that farmers do not have a long-term interest in these types of stewardship practices as they live in a dynamic environment in which urbanization processes have meant increasing pressure on land and water, and improved access to markets which has contributed to decreasing water quality. This lack of commitment is further encouraged by better income and job opportunities in other sectors. Although most peri-urban households still depend partly on agriculture, almost none of them see a prosperous future in it and the farm is mostly run by the older family members (most households consist of three generations). Children are advised to learn other (urban) professions for a better future (Evers, 2006), resulting at the same time in a loss of the local agricultural knowledge necessary for future food security and possible management of green productive space in the urban environment.

In general, wastewater is perceived in Hanoi as an inappropriate source of water for irrigation. In fact, many civil servants deny that it is even used for irrigation purposes. Therefore, it is not remarkable that farmers (and agriculture as a sector) are not regarded as stakeholders of the urban wastewater management network. At the same time, farmers have little knowledge of institutional aspects and proper agricultural practices with nutrient-rich irrigation water and show short-term spot market behaviour. The farmers also regard the quality of the irrigation water as poor, mainly because of its industrial constituents. Additionally, farmers prefer urban job opportunities above farming, because of income benefits and probably due to a preferential social status. These interrelationships do not contribute to long-term coherent development towards a sustainable integrated urban wastewater recycling system.

10.5 Towards implementation of international commitments on sustainable green cities and wastewater reuse

IWRM is not an objective in and of itself. It is a means to increase the efficiency and benefits of the scarce resources of water, land, finances and human capital and to convert them into usable products and services, without compromising the sustainability of the ecosystem (Rees, 2006). Urban IWRM approaches are necessary for realizing green cities. The process should however not focus too much on addressing priorities related to urban water and sanitation investments solely based on investments and the rates of return. Low rates of return have been common in many urban water management projects in cities in the developing world. This is reflected in poor utility performance and the lack of ability to set prices and collect revenue to recover operation and maintenance costs. Many of the benefits of investing in urban (waste) water management do not return to the water sector, but do flow back as improved health, welfare, environment and (agricultural) productivity. IWRM strategies need to consider low-cost technologies appropriate for local socio-economic and environmental conditions. Where public sewerage services are provided the question of cost recovery needs to be addressed. At the same time, decentralization and community participation likely yield only a few benefits without investments in skills and knowledge transfer.

As urbanization and population growth continues, water quality and food security will become major topics. Wastewater treatment, reuse and recovering the valuable elements will be an issue in all parts of the world. Currently in Hanoi the urban ecosystem is polluted because of a shortage of wastewater treatment capacity and a lack of knowledge on environmentally appropriate agricultural practices (shown by excessive use of fertilizers). As mentioned earlier, technologies are available for environmentally safe wastewater reuse, and there are good practices available globally (although mostly in developed countries). In the metropolizing areas of the developing world, it is not the lack of technologies for effective wastewater management that have resulted in the current poor state of affairs, but that the speed of urbanization and its negative consequences are outpacing the capacity of cities to adapt and to implement sustainable integrated water management projects.

The success of international commitments on sustainable development depends on local implementation. As described in Chapters 2 and 3 of this book various factors explain the process and outcome of policy implementation. Prominent literature suggests that the odds against implementation of policies increase with the number of decision points (Pressman and Wildavsky, 1973). O'Toole and Montjoy (1984) agree in general, but also argue the opposite that the probability of implementation increases with the number of units when the type of interdependence requires little coordination at the formative stage. Based on the hierarchical structure of the Vietnamese governance system (many decision points) and the complexity of the issue (requiring much coordination at the formative stage between units) one can assume difficulties are (would be) present in implementing integrated wastewater management plans in Hanoi. A more fragmented governance system has a higher likelihood of experiencing poor implementation. The complexity of the issue of sustainable green urban development, sanitation and agriculture not only demands a cooperative context, but implementation in countries in economic transition is also seriously limited by capacity. As with many other national governments, Vietnam has created a multitude of institutions (legislative and administrative) related to water at different levels and scales. This complex institutional network might work fine when well-coordinated, but the Hanoi case is an example where this coordination is lacking. A first step thus would be to set up a coordination body amongst the responsible institutions which links across levels and sectors (Evers *et al.*, 2010; Huibers *et al.*, 2010). Currently, in Hanoi (and most other metropolizing cities) agriculture, sanitation and urban development are separated. This fragmentation reflects the strong boundary between 'the urban' and 'the rural' domains. However, there have always been strong interconnections between agriculture and the city and new paradigms in urban planning demand integrated governance of both domains. These links can be clarified when water that is allocated from agriculture to urban use is returned to agriculture as valuable wastewater (Huibers *et al.*, 2010). This first of all requires negotiation amongst the diverse user groups, and probably also new appropriate legal and management frameworks. Coherence of the governance regime (levels, sectors, aspects, actors, policies, instruments etc.) is regarded as crucial for sustainable development of cities. Viewing the urban system from a holistic perspective wastewater can be regarded as a supply asset and wastewater irrigation as an environmental service. Wastewater use needs to be planned, investments made, health protection measures introduced and any public fears for the use of the waste effluent and the agricultural produce need to be addressed in combination with the protection of agricultural land in order for it to be a sustainable resource for agricultural production and a wastewater management system.

At the city level, the way that urban development is managed in terms of, for example, layout, density, building design and zoning can have significant implications for water service delivery and the costs involved. Options to manage demands, protect existing water sources or seize reuse opportunities may be restricted if land use planning, building or appliance regulation, school curricula, industrial development and pollution abatement, and ground water abstraction control remain uninfluenced by service providers. Further supply options could also arise if the new water cycle created by urbanization is viewed in its entirety.

Stakeholders have differing levels of access to resources such as funding, knowledge and power. Unless carefully managed, participation can reinforce these differences and risk further marginalizing the poor and other already marginalized social groups. The interests of unorganized or underrepresented groups will not be taken into account in decision making. The result of the decision-making process will be biased in favour of the more powerful stakeholders. For the urban poor the cost of engaging in consultation exercises can be significant as they struggle to meet their basic survival needs. As Simpson (2004, in Rees, 2006, p. 34) asked, 'when the rich can just turn on the tap, why do we expect the poor to engage in character building participation to get water?' In many urban contexts, direct participation may be unfeasible and unwanted. What is necessary are transparent, accountable and trusted mechanisms for representation.

Coherence and trust can only be realized through the involvement of a large and diffuse group of stakeholders. Molle and Hoanh (2009) remark that the Vietnamese Law on Water Resources does not mention the words 'participation' or 'participatory'. The main ministries responsible for water resource management, the Ministry of Agriculture and Rural Development and the Ministry of Natural Resources and Environment, compete rather than cooperate. In institutional terms, integration between urban wastewater management and peri-urban irrigation management is non-existent. There is no political context of consensus seeking and negotiated agreements for environmental policy implementation despite a context that has (the appearance of) mutual respect and trust. The Vietnamese use a more engineering and sector-based approach to combat environmental degradation. For urban farmers (often the poorer part of the urban population) direct participation may be a costly effort to get water, especially when they have something to lose at the same time. However, what is necessary is that transparent, accountable and trusted mechanisms for representation are in place (Rees, 2006).

Evers *et al.* (2010) described the behaviour of the actors in the fragmented governance system as spot market behaviour along the water (supply) chain. In such a fragmented system actors behave independently resulting in unpredictable and uncoordinated development of the system. Based on the short-term focus of the actors in the Hanoi system it is unlikely that this will result in long-term sustainable development. Evers *et al.* (2010) recommend a stepwise approach towards more vertical integration (coherence) in which the responsibility is at the government level since local farmers lack the capacity for substantial change. This is comparable to Bressers and Kuks (2005, p. 10), who state rightly that

> a change towards coherence occurs only when relevant actors acknowledge that coherence is necessary to prevent further deterioration of the resource and take action. Coherence is not a spontaneous development. Unlike an increase in complexity, then, developments in the direction of more coherence need some sort of deliberate attempt by motivated actors.

Based on the skills and capacities of the actors at different levels of decision making, decisions need to be made at the right level. This means that with little capacity present at lower levels, more centralized (top-down) detailed frameworks need to be in place. With high levels of capacity at the lower levels a more decentralized system can be present, though higher-level monitoring and regulation should still be in place to secure coherence in the larger system.

In the case of Hanoi, two essential elements (participation and coherence) for sustainable development of cities and the reuse of valuable waste products are largely missing. Hanoi is not an exceptional case, however, as is shown by the overall lack of progress on Millennium Development Goal 7 to provide proper sanitation to urban dwellers worldwide, the global informal use of wastewater in urban agriculture, the lack of recovery of valuable nutrients and the ongoing deterioration of urban and downstream environments (Corcoran *et al.*, 2010). In many situations a lack of capacity (money, knowledge, power to change) is also a very important factor hindering progress on sustainable development.

To integrate urban agriculture into urban wastewater management a fundamental rethinking of existing governance systems is required. Governance systems need to be able to improve urban sanitation, enable the recovery of valuable resources, halt environmental degradation and increase safe production of food in and around metropolizing cities. Large centralized sewerage systems (as they have been realized in developed countries) are very expensive to maintain in a sustainable manner The proposed combined urban water–waste–food management system needs to be accompanied with decentralization of sanitation and reuse systems and stakeholder participation (especially from farmers as users of the end-product and environmental service providers). Municipalities, like Hanoi, should thus be enablers of different mixtures of sanitation and reuse systems in new urban developments (possibly with private investments). Ideas from supply chain management and IWRM can help to optimize the management of such a wastewater–food production chain, coordinating the actions of the independent actors and paving the way for flexible but durable institutional frameworks for wastewater management. It follows that coherence within the governance regime is a central requirement for sustainable urban development. Recognizing the importance of urban agriculture for food production and green space, as described in many international commitments, urban planners need to identify and manage these land uses within the urban system. IWRM plans and participation of local users is however not a panacea for realizing sustainable systems. Understanding the possibilities of a holistic integrated approach that includes the transcending issue of urban wastewater irrigated agriculture can contribute to the development of green sustainable cities.

10.6 Conclusions

This chapter has addressed the question: why are international concepts and commitments on sustainable cities hardly transferred into integrated urban water resources management policies and practices in metropolizing cities in the developing world? Wastewater use in urban agriculture is a common reality. International commitments favour wastewater reuse development, as it is widely regarded as a solution to water scarcity as well as an environmental protection measure. The example of Hanoi depicts the complexity of the issue.

Contemporary technologies make it possible to reclaim wastewater for any purpose. The context, however, strongly influences the local choice and application of technology. Local contexts demand a smart combination of wastewater collection and treatment for the urban environment, based on existing infrastructure and institutions, financial capacity, space, availability of skilled personnel, and cultural perceptions on the use of wastewater.

For sustainable use of wastewater in (peri)urban agriculture cities need political decisions, clear institutional arrangements, financial incentives, policies and enforcement for the protection of agricultural land and to ensure the quality of the water resources. This requires special attention for wastewater that includes industrial waste and heavy metals. Ensuring agricultural land (or green belts) within the city boundaries is essential for safeguarding the land needed for wastewater use in agriculture or other low-tech land-based wastewater treatment systems and closing nutrient loops within the urban system. Securing land is a major challenge in rapid urbanizing areas where land is often designated to higher value purposes such as housing or industry, or informally inhabited by urban dwellers. Full cost recovery is desired when installing reuse systems. Farmers could even be paid as environmental managers which would better secure farmers' livelihoods and reduce the likelihood of them seeking non-agricultural jobs. Additionally, wastewater irrigation allows fresh water to be allocated to higher value uses. Treating waste as a resource (also in policy arrangement) opens opportunities for businesses to invest in services to transport sewage from households to farms.

IWRM offers valuable principles for the sustainable development of urban systems. However, the metropolizing context includes many parallel processes. Urban planning is often outpaced by the rapid urbanization process. Sustainable development of cities requires holistic perspectives much wider than IWRM. Sustainable development of urban wastewater management systems requires much bricolage and 'learning while doing' of transferring concepts about urban agriculture, sustainable wastewater management, recycling and recovering of nutrients, re-greening urban environments into existing urban governance systems and by doing so creating local context specific governance and management systems. Sustainable development requires creating coherence between all kinds of parallel development processes, but also space for these processes to develop. It requires balancing between open planning processes and the creation of a clear and hard framework for sustainable development. It is thus important to have a shared vision on sustainable urban development among the different stakeholders. At the same time it is as important to facilitate decentralized processes to contribute to sustainable development in developing local solutions. The art is not in trying to control it as one IWRM process, but to interactively implement parallel urban development processes in such a way that the parallel processes support and complement rather than frustrate and eliminate processes toward sustainable cities.

References

Barua, S. and Van Ast, J.A. (2011) 'Towards interactive flood management in Dhaka, Bangladesh', *Water Policy*, vol. 13, no. 5, pp. 693–716.

Biswas, A.K. (2004) 'Integrated water resources management: A reassessment', *Water International*, vol. 29, no. 2, pp. 248–256.

Bressers, H.T.A. and Kuks, S.M.M. (2002) *Integrated Water Management Regimes and More Sustainable Water Resources in Europe: A case study comparison*, University of Twente [European Research Project contracted by the EC-DG XII in respect of the Fifth Framework Programme 1998–2002], Enschede.

Bressers, H.T.A. and Kuks, S.M.M. (2003) 'What does governance mean? From conception to elaboration' in H.T.A. Bressers and W.A. Rosenbaum (eds)

Achieving Sustainable Development: The challenge of governance across social scales, Praeger, Westport, CT/London.

Bressers, H.T.A. and Kuks, S.M.M. (2005) 'Integrated regimes and sustainable use of natural resources: A multiple case study analysis', paper presented at the Conference on Sustainable Water Management: Comparing Perspectives from Australia, Europe and the United States, National Europe Centre, the Australian National University, 15–16 September, Canberra, Australia.

Bruinsma, J. (2011) 'The resources outlook: By how much do land, water and crop yields need to increase by 2050?' in P. Conforti (ed.) *Looking Ahead in World Food and Agriculture: Perspective to 2050*, FAO, Rome.

Corcoran, E., Nellemann, C., Baker, E., Bos, R., Osborn, D. and Savelli, H. (eds) (2010) *Sick Water? The central role of wastewater management in sustainable development – a rapid response assessment*, United Nations Environment Programme, UN-HABITAT, GRID-Arendal.

Drechsel, P., Scott, C.A., Raschid-Sally, L., Redwood, M. and Bahri, A. (eds) (2010) *Wastewater Irrigation and Health: Assessing and mitigating risk in low-income countries*, Earthscan, London.

EIU (Economist Intelligence Unit) (2011) *Asian Green City Index: Assessing the Environmental Performance of Asia's Major Cities*, Siemens AG, München.

Evers, J.G. (2006) 'Everybody's business is nobody's business: A study on the use wastewater in (peri-)urban irrigated agriculture in Hanoi, Vietnam', MSc thesis, Wageningen University and Research Centre, Wageningen.

Evers, J.G. (2011) 'Werk in Uitvoering. De toepassing van Interactieve Uitvoering in de praktijk', PhD thesis, University of Twente/CSTM, Enschede.

Evers, J.G., Huibers, F.P. and Van Vliet, B.J.M. (2010) 'Institutional aspects of integrating irrigation into urban wastewater management: The case of Hanoi, Vietnam', *Irrigation and Drainage*, no. 59, no. 3, pp. 336–344.

FAO (Food and Agriculture Organization) (2007) *Profitability and Sustainability of Urban and Peri-Urban Agriculture*, FAO, Rome.

Figuié, M. (2004) *Consumers' Perception of Tomato and Water Convolvulus Quality in Hanoi, Vietnam*, AVRDC – The World Vegetable Centre, Tainan.

Geldof, G.D. (2001) *Omgaan met Complexiteit bij Integraal Waterbeheer*, Tauw bv., Deventer.

GSO (General Statistics Office of Vietnam) (2009) *Report on Completed Census Results: The 1/4/2009 Population and Housing Census. Hanoi, Vietnam*, GSO, Hanoi.

GWP (Global Water Partnership) (2000) *Integrated Water Resources Management*, GWP, Stockholm.

Huibers, F.P., Redwood, M. and Raschid-Sally, L. (2010) 'Challenging conventional approaches to managing wastewater use in agriculture', in P. Drechsel, C.A. Scott, L. Raschid-Sally, M. Redwood and A. Bahri (eds) *Wastewater Irrigation and Health: Assessing and mitigating risk in low-income countries*, Earthscan, London.

Jiménez, B., Drechsel, P., Kone, D., Bahri, A., Raschid-Sally, L. and Qadir, M. (2010) 'General wastewater, sludge and excreta use situation', in P. Drechsel, C.A. Scott, L. Raschid-Sally, M. Redwood and A. Bahri (eds) *Wastewater Irrigation and Health: Assessing and mitigating risk in low-income countries*, Earthscan, London.

Luan, T.D. and Minh, N.H. (2005) *Institutional Aspects of Agriculture Management in Hanoi Periurban Area*, SEARUSYN, EU 5th Framework INCO2 funded research project, Hanoi, The Hague.

Molden, D.J. (ed.) (2007) *Water for Food, Water for Life: A Comprehensive Assessment of Water Management in Agriculture*, Earthscan, London/International Water Management Institute, Colombo.

Molle, F. and Hoanh, C.T. (2009) 'Implementing integrated river basin management: Lessons from the Red River Basin, Vietnam', IWMI Research Report 131, International Water Management Institute, Colombo.

O'Toole, L.J. and Montjoy, R.S. (1984) 'Interorganizational policy implementation: A theoretical perspective', *Public Administration Review*, vol. 44, no. 6, pp. 491–503.

Penning de Vries, F.W T., Acquai, H., Molden, D.J., Scherr, S., Valentin, C. and Cofie, O. (2002) 'Integrated land and water management for food and environmental security', Comprehensive Assessment Research Report 1, International Water Management Institute, Colombo and Global Environmental Facility, Washington, DC.

Pressman, J.L. and Wildavsky, A.B. (1973) *Implementation: How great expectations in Washington are dashed in Oakland; or, why it's amazing that federal programs work at all, this being a saga of the Economic Development Administration as told by two sympathetic observers who seek to build morals on a foundation of ruined hopes*, University of California Press, Berkeley, CA.

Ramírez, C.A. and Worrel, E. (2005) 'Feeding fossil fuels to the soil: An analysis of energy embedded and technological learning in the fertilizer industry', *Resources Conservation and Recycling*, vol. 46, no. 1, pp. 75–93.

Raschid-Sally, L. and Jayakody, P. (2008) 'Drivers and characteristics of wastewater agriculture in developing countries: Results from a global assessment', IWMI Research Report 127, International Water Management Institute, Colombo.

Raschid-Sally, L., Doan, D.T. and Abayawardana, S. (2004) 'National assessments on wastewater use in agriculture and an emerging typology: the Vietnam case study', in C. Scott, N.I. Faruqui and L. Raschid-Sally (eds) *Wastewater Use in Irrigated Agriculture: Confronting the Livelihood and Environmental Realities*, CABI Publishing, Wallingford.

Rees, J.A. (2006) 'Urban water and sanitation services: An IWRM approach', TEC Background Papers, no. 11, Global Water Partnership, Stockholm.

Sehring, J. (2009) 'Path dependencies and institutional bricolage in post-Soviet water governance', *Water Alternatives*, vol. 2, no. 1, pp. 61–81.

Simpson, R. (2004) 'How to increase customer orientation of public sector utilities', Presentation at Modes of Engagement with the Public Sector Water Supply Providers in Developing Countries, World Bank-Water Aid Workshop, 23–24August, London.

Smit, J. and Nasr, J. (1992) 'Urban agriculture for sustainable cities: Using wastes and idle land and water bodies as resources', *Environment and Urbanization*, vol. 4, no. 2, pp. 141–152.

UN-Water (2011) 'Challenges and opportunities for water in the transition to a green economy', *International UN-Water Conference: Water in the Green Economy in Practice: Towards Rio+20*, UN-Water, Zaragoza.

UNCSD (United Nations Conference on Sustainable Development) (2011) 'RIO 2012 Issues Briefs No. 5: Sustainable cities', United Nations Department of Economic and Social Affairs, New York.

United Nations (1992) *The Rio Declaration on Environment and Development*, UN Conference on Environment and Sustainable Development (the Earth Summit), Rio de Janeiro.

UNPF (United Nations Populations Fund) (2011) *The State of the World Population 2011*, UNPF, New York.

Van der Water, J. (2011) 'Change for the better? A study on changes of cultivation techniques in wastewater irrigated agriculture in peri-urban Hanoi, Vietnam', MSc thesis, Wageningen University, Wageningen.

Van den Berg, L.M., Van Wijk, M.S. and Hoi, P.V. (2003) 'The transformation of agriculture and rural life downstream of Hanoi', *Environment and Urbanization*, vol. 15, no.1, pp. 35–52.

Werner, C., Avedaño, V., Demsat, S., Eicher, I., Hernández, L., Jung, C., Kraus, S., Lacayo, I., Neupane, K., Rabiega, A. and Wafler, M. (eds) (2004) 'Ecosan: Closing the loop, Proceedings of the 2nd international symposium', 7 –11 April 2003, Lübeck, GTZ, Eschborn.

WHO (World Health Organization) (2006) *Guidelines for the Safe Use of Wastewater, Excreta and Greywater, vol 2: Wastewater Use in Agriculture*, WHO, Geneva.

11 Institutional innovation of water governance in Mexico

The case of Guadalupe basin near Mexico City

María Laura Franco-García, Lina Hendrawati Tan, Elena Carina Gutiérrez-Díaz, Cesar Casiano and Hans Bressers

11.1 Introduction

Various international organizations cite the need for water management to be organized on a 'river basin scale' and with a high degree of 'public participation'. These terms are associated with the universal remedies for water management problems that have developed in Chapter 2. However, context matters, and policies and concepts face specifics of place (Kotzebue, 2012). There are opportunities and hurdles associated with placing a new concept into the processes of water management implementation activities and planning into another country than where it was developed (Chapter 2). In this chapter we will study how these principles were implemented in a Mexican case.

The most significant factor affecting current changes to Mexico's water policy is the political opening-up and modernizing of Mexico's economy that began in the early 1990s. In 1989, Mexico's water institutions were strengthened with the creation of the National Water Commission (CONAGUA) and water policy focused on markets and efficiency as the country prepared to join the 1994 North American Free Trade Agreement. Only a few years later, Mexico signed several new free trade agreements, exposing agricultural producers and corporations to competition and economic pressure. This occurred at the same time that Mexico itself undertook great economic restructuring that included the privatization of state-owned companies, elimination of subsidies, initiation of market reforms, and a reduction in government size and capacity. Concurrently, Mexico was also influenced by the international water management paradigm reflected at the Earth Summit in Rio de Janeiro and the Dublin Principles (GWP, 2012), by a series of global water initiatives and by the World Bank's water resources policy and conditionality requirements. In 2003, the OECD released an Environmental Performance Review of Mexico

(OECD, 2006). It assessed Mexico's performance in achieving its domestic objectives and international commitments. The review provided, among others, the following recommendations on water management:

1 strengthen and further develop an integrated river basin approach to both improve water and forest resources management and provide environment-related services more efficiently;
2 reinforce current policies for raising water quality awareness and fostering public participation in river basin management; and
3 give greater weight to the protection of aquatic ecosystems (e.g. rivers, lakes, estuaries, deltas, wetlands) in water management decisions and activities.

In 2004, the 1992 (December) National Water Law was modified to implement the concept of integrated water management. In 1997, however, CONAGUA had already promoted the decentralization of water management since it divided the country into thirteen regional headquarters; each administered by a 'river basin governmental organization'. These 'organizations' provided the opportunity for regular citizens to be stakeholder representatives in 'river basin councils'. The Guadalupe river basin council is one such council. The 2004 law specifies the role, membership and activities of the Mexican River Basin Council (MBC) and indicates that the environment itself is a legitimate water user. Recent reforms to the National Water Law include the obligation for CONAGUA to generate participation amongst water users in the administration of the system and services. This same regulation obliges CONAGUA to consult the different actors involved, through the MBCs, in order to register their opinions and proposals in terms of planning, identifying problems and developing strategy for management, as well as evaluating the supply sources in accordance to the principles of sustainable development. There are guidelines for public participation described in the Rules of Organization and Operation of the MBCs (Comisión Nacional del Agua, 2004).

MBCs and their auxiliary commissions, river basin committees and groundwater committees (*cotas*) represent the institutional path for public participation and river basin management. The main purposes of their formation are (Comisión Nacional del Agua, 2004):

- increasing the level of communication among the stakeholders (MBC members);
- systematically checking the planning progress;
- disclosing results regularly;
- transparency in financial provision and distribution;
- gaining visibility in the media;
- regularly involving citizens with some level of authority in the communities;

- increasing the awareness and environmental education within communities;
- implementing programmes to actively stimulate participation;
- establishing an atmosphere of equality among MBC members;
- intellectual stimulation and feedback to the council members from an expert panel to guarantee constant improvement; and
- introducing capacity-building programmes for MBC members with a focus on management skills and leadership.

These purposes indirectly serve to improve the likelihood of more sustainable water resources in three respects: ecological, economic and social.

The different contexts and development histories of the various MBCs increase the difficulty in comparing the level of their effectiveness. Thus, we have chosen in the limited scope of this chapter to perform instead a single case study that is relevant because it was initiated in a bottom-up fashion while other MBCs have been initiated by CONAGUA (national governmental level). This MBC also has a generally good reputation. Even a relatively successful case will, however, be required to face some difficulties that are typical for the transfer of integrated water management concepts. A relatively successful case can teach us about both the various difficulties and the opportunities of transferring international water management concepts into the Mexican context. The river basin scale principle can create tensions with existing jurisdictional scales and public participation can shake up existing power relations and competence domains (see also Chapter 2). These approaches are potentially difficult to implement as part of the Mexican context, even when their underlying values are generally supported.

The central research question of this chapter is: What contextual factors have been positive or negative conditions for the efficacy of Guadalupe's MBC, interpreted as an example of the transfer of international water management concepts into a specific Mexican context? This is an empirically based chapter that is supported by the theoretical framework provided in Chapter 3 in reference to the Contextual Interaction Theory (CIT) model and to the four qualities of a supportive governance regime, as we specify below when mentioning the following sections. The basis of our empirical analysis is formed by the following sources: participative observation by one of the authors, master thesis research with document analysis and interviews by another author (Tan, 2010), and second rounds in 2012 of document analysis and interviews with most members of the council (stakeholder representatives), often in the presence of some of the sector members that they represent. The interviews had a semi-structured nature with questions that largely follow the structure of this chapter. With one exception (the president of the council) we quote our interviewees in such a way that they cannot be identified individually.

Section 11.2 describes the background of the case. Thereafter, we frame our research question by using CIT components which are provided by Chapter 3. We assess in Section 11.3 the accomplishments of the MBC and the relevant general external context observed to be present in 2012. This is followed by a more detailed explanation in terms of the consecutive layers of context. In Section 11.4, the extent of the constellation of stakeholders involved, the degree of cooperation within the council and the most crucial issues in their deliberations are dealt with as a first order of explanation. In terms of the Contextual Interaction Theory (Chapter 3), this section describes the interaction process. The next layer of explanation is explored in Section 11.5 by elaborating on the motivations, cognitions and resources that the stakeholders bring into the process. Questions related to the specific actor characteristics were not directly asked in the interviews. This information was derived from observations and several questions in the interview rounds. In Section 11.6, the four qualities of the structural context of the governance regime (extent, coherence, flexibility and intensity) are used to complete the analysis. Finally, Section 11.7 provides the general conclusion that has been reached based on the empirical analysis.

11.2 Guadalupe basin and its MBC

The Guadalupe river basin is located in the northern part of Mexico City, though it belongs to the State of Mexico. Mexico City has been merged with a number of other municipalities within the State of Mexico and the resulting agglomeration is called the 'Metropolitan Zone of the Mexican Valley'. It represents the most populated area in Mexico with over twenty million inhabitants (Consejo Estatal de Población, 2010). Guadalupe lake itself is located in Cuautitlán Izcalli municipality, but the whole river basin consists of five municipalities: Atizapán de Zaragoza, Cuautitlán Izcalli, Isidro Fabela, Jilotzingo and Nicolás Romero (GRAVAMEX-SC, 2004).

In May 2004, the first massive incident of fish deaths occurred in the Guadalupe lake. This was followed by two similar mortality events, first in December of the same year and the second in February 2005. In total 31.5 tons of dead fish were found (Gutiérrez, 2006). Just after the first event, two meetings were held in June 2004 by members of organizations related to the Guadalupe's river basin. The organizations were water and ecology departments of the municipalities of Nicolás Romero, Atizapán de Zaragoza and Cuautitlán Izcalli, the Water Commission of Mexican State, the National Water Commission, the National Environmental Protection Attorney Agency, Secretary of Environment, Government of State of Mexico and specialists from the department of Limnology, Toxicology and Fish Histopathology of the National Autonomous University of Mexico (UNAM). They produced a public declaration about the following commonly agreed upon items: (1) the need to establish a diagnosis of the sub-tributaries of the Guadalupe river basin; and (2) the relevance of monitoring the physical, chemical and biological water properties in the Guadalupe river basin. The process to investigate the causes of the incident and produce the public declaration started in September 2004 and ended in January 2006.

The resulting monitoring activity was conducted with the cooperation of the municipalities. This was mostly supported by the municipality of Cuautitlán Izcalli. Water quality (physical-chemical and biological parameters) and histopathology fish analyses were carried out at UNAM facilities. During the initial meetings, the discussions concentrated on how to solve the problem as expediently as possible. The Cuautitlán Izcalli municipality was the most active one in working to stop the occurence of mass mortality of fish since the river basin is located within its municipal area. Directly following the meeting, all stakeholders agreed that action needed to be taken so that such a problem would not happen again in the future. In May 2005, one year after the first massive occurrence of fish deaths, the main stakeholders from different sectors decided to establish a river basin commission under the call of the Cuautitlán Izcalli representatives and supported by the diagnosis of the Guadalupe's river basin problem. The official 'Guadalupe River Basin Commission' (CCPG) was created in January 2006 under the framework of the National Water Law and as an auxiliary organization of the Council of Valle de México river basin (Comisión Nacional del Agua, 2011).

The government stakeholders represented in the CCPG are the mayors of the five municipalities, and regional and federal level governmental representatives who have responsibilities regarding aspects of water bodies. In the CCPG structure, there are also societal representatives (educational institutions, NGOs) and those from the private sector. In 2012 the majority of the members were governmental representatives, which is a larger share than the law envisions (maximum 50 per cent of governmental organizations). The commission itself planned to involve more stakeholders from the social sector.

A new mass mortality of fish has not occurred recently, but the commission continues to work to solve other problems and to prevent any environmental impacts that would develop as a consequence of insufficient basin management. One of the problems that continues to be discussed during the CCPG's meetings is the level of the drinking water supply, which is insufficient to meet current demands. They concluded that this particular problem is caused by the low water quality in the Guadalupe river basin. The sewage of the residential districts in the neighbourhood, as in so many developing countries (Trivedi, 2006), is being discharged to the rivers directly without treatment and hence severely reduces the water quality in the Guadalupe basin. The problematic discharge occurs due to the lack of municipal infrastructure such as separated drainage for sewage and rainwater; and collection, treatment and landfills for solid urban waste. The Guadalupe river basin is the final discharge point for several rivers containing wastewater from human settlements within the five municipalities

(Gutiérrez and González, 2005). It is estimated by CONAGUA discharges sum a total volume of 14,924,455 m^3 per year of untreated wastewater originated from Nicolás Romero (90.7 per cent), Atizapán de Zaragoza (6.6 per cent) and Cuautitlán Izcalli (2.7 per cent), through the rivers Cuautitlán, San Pedro, Xinté and Tecuane. Additionally, the solid waste disposed into illegal dumping sites instead of the legal landfills further reduces the water quality in the basin. This problem is worsening given that the municipality of Cuautitlán Izcalli, in its continuing programme of solid waste removal, is each year only capable of removing a limited volume of the plastic polyethylene terephthalate (PET), tyres and garbage floating on the river towards the Guadalupe basin.

The diagnosis shows that the intense growth and sprawl of urban settlements in the basin area are the main cause of the existing poor conditions and increasing deterioration of the river basin (Gutiérrez and González, 2005). Ninety per cent of the approximately 355,000 inhabitants are located near the river basin, mostly in Nicolás Romero, which is predicted by the CCPG to increase to over 425,000 inhabitants by 2020. The consequent impact on wastewater generation and discharge without any treatment is estimated to be nearly 18,000,000 m^3 (Gutiérrez and González, 2005), assuming the same amount of water supply. Another important problem related to the people living at higher geographical positions in the basin is deforestation. The availability of water is in turn affected by this deforestation that is caused by changes in land use. Forest is turned into agricultural land and agricultural areas are developed into urban spaces, which has been happening at an accelerated pace to allow more settlements. This issue has also become one of the concerns of the municipalities therefore limiting urbanization and deforestation has been included in the municipal development plans.

In 2006, members of the CCPG collaborated with CONAGUA and the Mexican Institute of Water Technology (IMTA) in order to develop a strategic plan to address the river basin problems in an integrated approach (IMTA, 2006). The specific goals of the *Programa Hídrico de Gran Visión de la Cuenca a Presa Guadalupe* (Desarrollo y sistemas, 2008) are:

1 promote river basin sanitation
2 achieve sustainable water utilization (drinking water, fisheries, irrigation, tourism and recreation are among the most important)
3 accomplish suitable natural resources management
4 support the development of rural areas
5 promote public participation
6 execute governmental, institutional and financing coordination
7 ensure the implementation of the legal framework

From those seven objectives, seventy-seven strategic actions have been elaborated and are planned to be executed before 2030. The strategic plan is framed as being part of the National Hydrological Programme for 2007–2012 (Comisión Nacional del Agua, 2008).

In the first round of interviews, which were conducted in 2009, it was found that there is some difference between the motivations of the governmental actors (implementing the water policies) and that of the user groups (feeling of responsibility and commitment to the environment and society). However, these motivations are not necessarily contradictory. Furthermore, a chairman with an interest in coordinating the CCPG actions towards success was bridging the communication between the other two groups. In addition, all the stakeholders involved shared the willingness to be collaborative and this is a key factor in achieving the goals of the strategic plan. This includes the willingness to be actively involved and participate in this commission, actively sharing information that might be useful for the basin area, and having the enthusiasm to reach consensus in the commission. The fact that the development of the commission was not urged by the new law, but rather developed through bottom-up mechanisms due to local problem pressures and was thereafter adopted into the law's framework, helped to achieve such consensus orientation. Although, consensus orientation is not seen as typical for Mexican society (Barkin, 2006).

11.3 Awareness and action: what has Guadalupe's MBC achieved?

Since 1943, water from the river basin has been used for the irrigation of agricultural areas located in the states of Hidalgo and State of Mexico. However, developing alongside the high level of population growth in Mexico City, the water in the reservoir is becoming mainly used for recreational purposes. The council is organized by 'working groups' and the particular working group in charge of the water projects is responsible for monitoring the programmes from higher-level authorities such as those established by the 'Commission of Water in State of Mexico'. This working group decided to construct sewage collection infrastructure for the populations which were settled around the river basin. A hydrological plan was developed in order to establish the general problematic of the area and specifically the role therein of the water resources. Environmental audits began under the terms of the environmental agency at Mexico State level with a special focus on municipal water use. As for the waste management, campaigns for PET collection were initiated, one in the Cuautitlán Izcalli. Regarding education, several agreements were signed to develop more collaboration with schools. Two books were published, one about vegetation and the second about migratory bird species. The forestry working group was successful in developing a pest control programme. Projects to improve the cultivation of mushrooms were started and led by a local teacher and the commission's coordinator. All the working groups aim to train and educate the general public by raising awareness about the environmental problems faced by the five municipalities involved in the CCPG.

In light of these developments, it is thus notable that the water quality as such has not sufficiently improved. Studies performed by UNAM and the IMTA show that the pollution level is still high. The presence of *Microcystis* algae provoked the lack of oxygen that led to the massive fish deaths mentioned earlier in this chapter. These studies were performed from 2004 to 2006 (Gutiérrez and González, 2005). The achievement of having an operational management agency is important because of its role in coordinating all the sectors and the value of having someone for coordination of projects and actions that can have a longer-term presence. The operational management agency is subsidized with federal and state resources with a one-off grant for establishing the agency.

Despite the lack of measurable improvement in water quality since the establishment of the CCPG in 2006, some improvements have been measured in terms of environmental quality conditions of the river basin ecosystem. The strategic plan described in the previous section has been executed from the basis of a clear diagnosis for each of the three regions within the basin. The main purposes of the strategic plan are: improving ecological conditions; preserving ecosystems; and raising awareness in order to encourage public participation towards sustainability of the basin. Apart from the plan itself, it is quite difficult to assess precisely the surplus value of the CCPG because investments were undertaken by the separate governments and could not be accomplished by the commission itself. Understandably, the governments attribute the achievement of these measures to themselves rather than to the commission and it is hardly possible to know whether these measures would have been taken without the CCPG and its strategic plan. Most of the efforts have been incorporated into environmental awareness programmes in the basin, though some of the governmental members maintain that more investments should be made into the development of projects for wastewater treatment plants upstream of the Guadalupe river basin. Some investments were made for facilities improvement after the creation of the CCPG but the State of Mexico denies that these investments have been made in order to meet the CCPG's goals (Tan, 2010). Also the number of projects that incorporate payment for ecological services has increased in latter years.

The farmers, including those working in aquaculture, are positive about the changes the commission has brought about. The rules for aquaculture have become clearer and also the measuring of water quality and actions to prevent pollution are now seen as assets to this sector. Farmers from the upstream region observe that most of the investments have been made in the downstream region of the basin. This sector has taken its own initiatives and has organized campaigns for litter collection in the countryside and forests. They have also cleaned up some clandestine dumping areas and protected the aquaculture facilities where some visitors had been polluting the trout ponds. The sector has also, for instance, invested in tree-saving projects such as lime application for parasite control. All in all, various ecosystem-related activities have been undertaken but some representatives of the social groups do not attribute these to the CCPG.

The representatives of social groups see public visibility as a problem: 'The impacts have been perceived as low level by the society. We have a plan, environmental monitoring, agreements signed with authorities but what is visible is still the water pollution problem. The population does not see any facilities under construction to solve the real pollution problems. Therefore it would be convenient to explain and communicate to the population about our administrative progress and the expected work in the medium and long term.'

Although there is still much to do in terms of water quality improvements (pollution is still introduced and transported all along the water trajectory), the attention for water management as such has substantially improved. This also benefits recreation, leisure and ecosystem services. In any case, what is frustrating for some of the CCPG members is that CCPG has not yet managed to obtain resources from upper governmental levels. As a result, they are not able to carry out large projects on their own. Some projects are carried out by the municipalities. They have made certain investments and have been able to construct a wastewater treatment plant along the Xinté river, which has not yet been operated because they have not acquired sufficient resources to install the necessary electricity infrastructure (280,000–350,000 US dollars). They also funded some river sanitation programmes which have supported the cleaning up of the waters and shores. These actions were also not seen to formally result from CCPG's initiatives. The total federal government financial contribution totals 70,000 US dollars, which is the same amount that was contributed by the government of the State of Mexico. Other contributions from CCPG member-institutions have also been made although the amounts are unknown.

CCPG's president sees the commission mainly as auxiliary to supporting the policies and strategies of the state government. Before the formation of the commission, only governments were involved in water management, but now a much wider array of civil groups participate and the commission has improved their relationships. The CCPG's president states:

> The main goal as a group of people is to support change towards more of an orientation directed to the quality of life, to create commitment and a feeling to take care for the landscape. Earlier some of the stakeholders were really killing the place. Now they are better aware of the value of the river basin. So, the results are not just about water quality. It is also about educating young people, also at schools and universities.

Therefore, we can conclude that the commission has concentrated on monitoring, planning, consensus building and raising awareness. Water management activities increased, though not yet to a sufficient level to attain the goals of the strategic plan. It was observed that most of the CCPG stakeholders gave signals of frustration due to the slow pace of progress. Furthermore, as the CCPG and its chair can only encourage others to take measures and their own capacity to act is very limited, it is difficult to assess whether such activities would also have taken place without the commission.

11.4 The actor constellation and the nature of the interaction process

In order to assess how the commission functions, it is important to see whether all relevant stakeholders and their corresponding issues and interests are included. Reaching consensus in a cordial atmosphere is seen as more likely to occur when only a subset of actors is involved and others are excluded, although such an approach would not really be inclusive. The array of stakeholders involved is in this case, however, rather large and has also evolved over time. There are some dynamics related to the inclusion of new stakeholders when they prove to be important for achieving the solution. In the last eight years both the fishing community as well as the people living in the forest have been involved. Furthermore, it was suggested by one of the social group representatives that because new issues continue to arise, new groups could still be drawn into the commission. As such, a social representative of CCPG mentioned that the tourism sector, schools, interested individuals and also deputies in the state parliament could be possible future relevant additional groups.

While we can label the composition of the actor constellation as relatively open to those who want to participate, this does not mean that all actors are equally committed to doing so. For instance the municipal representatives (all mayors) tend not to participate. They have experienced a continuous need for more resources and support and are often simply too busy to attend; some interviewees reported that meetings are often organized on rather short notice. Furthermore, urban development programmes can work from a separate guiding vision than that of the river basin upgrading perspectives. For both societal and governmental representatives this low level of participation need not be a problem however. One interviewee stated: 'Though the mayors are mostly not present, that is no problem as long as they support the action when needed. There are other contacts with them than just the council meetings.' Nevertheless, the dynamics related to unequal participation enable to some extent the dominance of visions and responsibilities of a few stakeholders.

All interviewees report that there is a high degree of openness regarding the issues that are debated in the meetings. The most sensitive subject is the discussion on the use of financial resources. Should they for instance concentrate efforts on developing a new ecological plan or on educational issues? Another difficult issue has always been how to identify who is responsible for a certain instance or type of water pollution and as well who should pay for its remediation. There are also other issues related to urban developments that are not part of the private sector's area of competence. One interviewee states:

> Subjects are openly presented and 'treated' at the right time. The problem is that some decisions are not taken in the plenary sessions. Sometimes one of the working groups seems to provide solutions without taking into account the other members' opinions.

At meetings though everybody can express their opinions. 'In general terms there was consensus within the working groups and afterwards in the plenary sessions where the proposals were presented and voted upon.'

The monthly meetings have an atmosphere that was reported by all as being very cordial and collaborative. Members are highly committed to realizing their responsibilities within the basin. The meetings include positive messages about the commitment of all of the partners and exchanges of experiences, and consensus is often achieved among the various opinions following the discussions. There is a general feeling of openness between most of the CCPG actors. The issues that they discuss and debate concern substantial resources and importance, yet despite this, achieving commitment across the territorial governments can be challenging. The nature of the interaction process is, however, generally quite collaborative.

11.5 The characteristics of the stakeholders involved

A collaborative nature for an interaction process is often a necessary condition for results, though not always a sufficient condition because of the dependency on others outside the process for the resources needed to implement the proposed activities (Owens, 2008). According to the concepts of the Contextual Interaction Theory as elaborated upon in Chapter 3, the nature of the interaction process is understood by analysing the motivations, cognitions and resources of the actors involved.

At a general level, *motivations* of the involved actors were observed to align quite well. Although this is understood to have led to the collaborative process, this does not mean that individual stakeholders' motivations are similar. On the contrary, examining the statements made by the interviewed actors reveals a very wide array of core motivations. The range includes finding means to obtain more certainty about present and future rules as well as generating more public attention and respect for another sector's positive role. Furthermore, some actors intensively aim to explore the potential for tourism development. One of the most prominent motivations was the preservation and promotion of ecological quality through improvement of sanitation measures. A further motivation highlighted in the interviews was to develop a more integrative policy planning. Its purpose should be to align protection and development measures, and to promote land use planning to prevent uncontrolled development. A final motivation that was mentioned related to undertaking a bottom-up approach to pollution prevention by educating people in ecological responsibility. This great variation of key motivations is quite interesting and particularly so based on the recognition of the alignment that is reached in the process. In other words, the motivations are different, but not contradictory.

In terms of *cognitions*, there is a variety of ways to look at the functions of the basin and the role of the commission, though there is more overlap between the various stakeholders than was the case with motivations. Many actors emphasized the importance of the various natural functions of the basin, such as healthy forests capturing rainfall for water supply, clean water to support ecological and agricultural functions, and avoiding flooding. Other actors emphasized that there is a big difference between the problem perception in the cleaner and less populated part of the basin where the water is captured in a quite natural area and the densely populated part where the waters are polluted and that receives almost all media attention. Each municipality involved has a different geographical, and consequently sometimes political, position in the basin. Others add the observation that social groups and individuals are still underrepresented and too unaware of the problematic to be able to function as a source of permanent pressure to enhance the process. Improving public participation and 'making people understand that sustainability is also a way of living' is thus essential. Again, these different emphases did not harm the collaborative process, as they were followed by gradual mutual recognition of each other's points of view.

The relevant *resources* at stake in the process were observed to be of a very mixed nature; sometimes positive, sometimes negative. The lack of strong social pressure to support the CCPG goals was mentioned as being not supportive of the interaction processes in the committee. For the stakeholders themselves, the lack of resources, time and financial compensation for their efforts also plays a role. Regarding knowledge, the link with universities is mentioned as a supportive aspect for the collaboration. The lack of funding was indicated as being a hindrance to the CCPG's operation by most of the CCPG members although their perception was not fully negative because some examples of resources being applied to projects were mentioned as very positive. The sectors can also sometimes take measures that are not necessarily costly, such as protecting the forest or making sure that the re-discharged water from fish ponds obtains sufficient aeration to allow for natural purification. However, the projects rely mainly on 'external' support for attaining the necessary financial resources. External is here placed in quotation marks, because part of it should come from stakeholders that are formally participating in the commission, but who separate their decision-making process related to those grants from their participation in the process. Both municipalities and Mexico State's government do contribute. However, this is primarily done in the context of supporting their own or national policies. This is even more true for the project support that in principle can be sought at the national and international level. Some legal complexities were also discerned as being influential. The commission now has the required legal stature that enables it to obtain national grants, yet mixing public and private support however seeking international grants is still problematic.

Motivations and cognitions among stakeholders are observed not to be identical but of the type that can be aligned during the process. Obtaining the resources needed is a potential 'hot issue', but as much of the resources need to be drawn in from other decision-making arenas, this has not caused deep cracks in the collaborative nature of the process. This can be seen as positive because the structure of the council and goodwill among its members survived despite the low level of achievements and concrete results.

11.6 The qualities of the Mexican governance context

The CCPG operates in a governance context that includes organizations, policies and rules at various administrative levels. According to Chapter 3 these should have a sufficient extent so that all relevant subjects are dealt with, have sufficient coherence so that they do not contradict each other and instead create synergies where possible, be flexible enough to allow for projects and measures that are adaptive to unforeseen opportunities as they arise, and intense enough to create a momentum for change towards a better status of the basin.

Concerning the *extent*, there were not many remarks made by the interviewees. The actors in the aquaculture sector feel that they are overlooked as an economic sector and consequently do not obtain support for their plans. Forestry representatives have a similar feeling. Both attribute this partially to the under-coverage of their sectors in the media. An interviewee stated as an example that news programmes prefer to pay attention to 10,000 new trees planted near one of the big population centres than to the 60,000 that are yearly planted by the people in the forest of the high-basin region by their own initiative. This situation has had some demotivating effects for some of the CCPG stakeholders. The support given by the governance regime does not include these sectors while they are in fact equally important for the status of the water basin.

Several issues were observed to negatively influence the *coherence* of the regime. The linkages between government levels and between the public and private domain are not clear for the societal representatives of the CCPG. One interviewee said:

> The National Water Law states that the responsibilities for water management are at federal and state levels which in practice does not

> really happen because at the local level the water committees allow for the private administration of water. Under these circumstances, water is no longer a common good and instead it becomes a commercial good. This situation is, in my opinion, a mistake of the National Water Law application and it has been discussed within our CCPG but we do not have the jurisdiction to penalize these actions.

The National Water Law was reformed in 1992 to enable the private sector to participate in the management of water. This developed as a response to the reforms in 1983 when water management responsibility was assigned to the municipalities and many of them declared not to have the capacity to handle it (Barkin, 2006). Another interviewee added: 'The laws do not link with the problematic at the municipal level. The laws seem to be too broad to be able to be applied to every stakeholder in the basin and that makes them lack intensity during implementation.' The link between policy and implementation is also low: 'The government does not charge the companies which pollute. The law is not implemented because it is not checked that who pollutes pays in return to restore the negative effects.' More generally, the same societal CCPG representatives also indicated that:

> It can be contradictory that each level is looking from its own purpose. A mayor might want concrete action now. The federal government takes the whole picture into account and thinks along broader lines and longer terms. Even among the five municipalities, there are different emphases. It is not so much a difference between positive or negative, but more a matter of timing. One of our most important achievements is the long-term plan, consisting of vision and actions.

So while the federal vision is broad, but not directly applicable at lower levels, there is a lot of fragmentation. A main virtue of the commission is to collect different interests back together into its visionary plan, even though it still depends on a variety of authorities with different outlooks for its realization.

The external dependency for support is also reflected in the assessment of *flexibility*. This is understood here as the freedom of the commission to seek appropriate solutions that maximize the use of temporal or spatial opportunities and create synergies where possible. One of the societal representatives stated that: 'There is not all the freedom that it is required, most of the limitations are around the economic resources needed to invest in the CCPG programmes', while the governmental stakeholder perception was explained as: 'They allocate resources for water management issues but there might be mismatches with the actions planned, like giving attention to different types of pollution and degradation factors.' Yet another governmental representative provided the following perspective:

> Some of the proposals come from higher governmental levels – top down – but do not always suit the CCPG's needs. For instance, some programmes coordinated by the 'Water Commission of State of Mexico' could be implemented for the Guadalupe river basin but do not really match the Guadalupe basin's needs.

Thus, the fragmentation of the regime also leads to inflexibilities. Support can only be obtained for certain objectives while these might not be at the core of the matter for the integrated management of the Guadalupe's river basin.

Given these findings, it follows that the *intensity* of the governance regime is experienced as low. When asked whether the support for change was sufficient, one interviewee was very clear in saying: 'Of course not, resources are short, both the economic and the administrative ones.' Federal ambitions may be high, but they do not often provide sufficient means to carry them out. When it comes to actual support, the commission depends on a wide array of other governmental levels and scales. The commission itself is only minimally supported with staff and money. Furthermore, the relations with those governments need to be re-established over and over again. A positive perspective was given by one of the interviewees:

> Since the start eight years ago we had two presidential elections and five local elections. We had to work with fifteen different mayors. Thus, resources need to come from many different people. Almost all were positive. Some are actually learning with us. In a sense we educate them.

While the intensity of the regime is still low, hope is still present for future improvements.

11.7 Fertile ground or hostile environment? A conclusion

The overall picture that arises from the above analysis shows a governance context that is only in principle supportive. It does not adequately equip organizations like the basin commission with sufficient support for its integrated ambitions for the basin nor does it give them the legal rights to collect those resources themselves. Instead, the commission depends on mid-level governments that view each project from the perspective of how it can address their own policies and are themselves only to some extent committed to the process. With these sharp-edged issues being partially excluded from the commission's interaction process, the variety of motivations and cognitions of all the stakeholders in the group were able to be bridged and some joint feeling of responsibility, mutual trust and awareness of the basin's problems developed. The commission has also tried to be very open and transparent to the press and to the people that the various stakeholders are being represented. This prevented the development of a culture of distrust and adversarial politics within the commission itself. The 'few achieved results' related to raising awareness both about the ecological status of the basin and about the necessity of joint action to generate more concrete results in the implementation of water management activities.

The perception of success depends on the actors' expectations. This supports the perspective of the commission's chair:

> Ultimately, it is about a better quality of life. A better quality of water, forest and landscape serve that goal. These parts are interconnected. Thus, we should work at each part and place, under an overall vision. The vision and concrete projects are mutually related. We started with drafting an initial plan. That took two years. It is both a blueprint and a vision. It is concrete, but should also be adaptive, because you need the people for any action.

The central question of this chapter was stated as: *What contextual factors have been positive or negative conditions for the efficacy of Guadalupe's MBC, interpreted as an example of the transfer of international water management concepts into a specific Mexican context?* How would the existing Mexican governance system be able to deal with organizations with river basin boundaries, and with extensive public participation? In both cases the answer is that the integration of these two concepts is still in progress. The CCPG chairman noted that in the beginning there were even concerns about whether or not the process would continue, because of the lack of cooperation by the municipalities. Later some young people that were educated by him and his university became mayors. That was helpful to get more respect for the committee and a greater understanding of the importance of the process. While river basin management and public participation are realized to a large extent and in a productive way in the commission itself, the integration of the commission's process into the mainstream policy process will continue to be an ongoing process for some years to come.

References

Barkin, D. (coord.) (2006) *La gestión del agua urbana en México: retos, debates y bienestar (The urban water management in Mexico: challenges, debates and well-being)*, Universidad de Guadalajara, México.

Comisión Nacional del Agua (2011) *Estadísticas del Agua en México (Statistics of water in Mexico)*, Edición 2011, CONAGUA, SEMARNAT, México.

Comisión Nacional del Agua (2008) *Programa Nacional Hídrico 2007–2012 (National Hydrological Programme 2007–2012)*, CONAGUA, SEMARNAT, México.

Comisión Nacional del Agua (2004) *Ley de Aguas Nacionales y su Reglamento (Law of National Waters and its Regulations)*, CONAGUA, México.

Consejo Estatal de Población (2010) 'Zonas Metropolitanas del Estado de México' *(Urban Zones of Mexico State)*, online, available at: www.edomex.gob.mx/coespo/

indicadoressociodemograficos/zonasmetropolitanas/index.htm, accessed July 2011.

Desarrollo y Sistemas, S.A. (2008) *Programa Hídrico de Gran Visión de la Comisión de Cuenca Presa Guadalupe (Hydrological Vision Programme of Guadalupe's River Basin),* CONAGUA, México.

GRAVAMEX-SC (Gerencia Regional XIII de Aguas del Valle de México y Sistema Cutzamala) (2004) *Compendio del Agua (Water Compendium),* México.

GWP (Global Water Partnership) (2012) 'Dublin Rio Principles', online, available at: www.gwp.org/The-Challenge/What-is-IWRM/Dublin-Rio-Principles, accessed 12 June 2012.

Gutiérrez, D.E. (2006) *Antecedentes, perfil y estructura de la Comisión de Cuenca Presa Guadalupe (Background, profile and structure of the Guadalupe's river basin),* Documento de trabajo, H. Ayuntamiento de Cuautitlán Izcalli, México.

Gutiérrez, D.E. and González, M.T. (2005) *Diagnóstico de la Subcuenca Tributaria Afluentes de la Presa Guadalupe (Diagnosis of the tributary streams of Guadalupe's river basin),* Documento de trabajo, H. Ayuntamiento de Cuautitlán Izcalli, México.

Hendrawati Tan, L. (2010) 'Stakeholder analysis with a Dutch conceptual framework for integrated water management in a Mexican case', master's thesis, University of Twente, Netherlands.

IMTA (Instituto Mexicano de Tecnología del Agua) (2006) *Plan Estratégico Hídrico de la Cuenca Presa Guadalupe (Strategic hydrological planning of Guadalupe's river basin),* CONAGUA, México.

Kotzebue, J.R. (2012) 'Spatial misfits in multi-level governance: Impacts on the small island state of Malta', PhD thesis, University of Twente at Enschede, Netherlands.

OECD (2006) *Water Management Policy Brief for the Mexican Project by Environment Directorate,* OECD, Paris.

Owens, K.A. (2008) *Understanding How Actors Influence Policy Implementation: A comparative study of wetland restorations in New Jersey, Oregon, The Netherlands and Finland,* Enschede: The University of Twente.

Trivedi, R.C. (2006) 'State of water quality and its management in India', in V.I. Grover (ed.) *Water: Global common and global problems,* Science Publishers, Enfield, NH.

Part III

Transferring an emerging concept

12 Translating the global climate change discourse to the local

An analysis of Dutch storylines on adaptation

Maya van den Berg

12.1 Introduction

It is increasingly being observed that climatic changes are taking place within all of the world's continents (Parry *et al.*, 2007). Therefore, in addition to the widespread attempts to curb the human-influenced emissions of greenhouse gases through mitigation, there is an increasing need for adaptation to the inevitable consequences of climate change. These consequences include many water-related impacts such as rising sea levels, greater variation in precipitation patterns and increasing flooding risks.

Climate change adaptation (hereafter 'adaptation') has increasingly gained attention since the United Nations Framework Convention on Climate Change (UNFCCC) introduced it to the global climate change discourse in the mid-1980s (Schipper, 2006). Adaptation is commonly defined as 'adjustment in ecological, social, or economic systems in response to actual or expected climatic stimuli and their effects or impacts' (Smit and Pilifosofa, 2001, p. 881). Adaptation includes both actions towards adaptive capacity so that actors can (better) adjust to changes and, as it has been understood in this study, towards the implementation of adaptation decisions (Adger *et al.*, 2005).

This chapter deals with how the concept of adaptation is interpreted in policies and how these interpretations are being shaped by their context. Clearly, incentives for adaptation can vary between different governmental layers. While national-level governments might be encouraged towards actions by an international policy concern from the global discourse, the signal at the local level might stem from a national policy providing rules or incentives to include adaptation as a local policy concern (Nilsson *et al.*, 2012).

While introduced at the global level, adaptation to the impacts of climate change is not primarily being organized at the global scale (Adger, 2001). Adaptation is commonly seen as a local issue since climate-induced impacts are felt at the local level (e.g. Wilbanks and Kates, 1999; Wilson, 2006; Storbjörk, 2010). Accordingly, many of the corresponding responses are found too at the local level (Cutter, 2003). However, many local decisions are shaped by interactions at higher levels that may mandate, encourage and inform actions (Wilbanks and Kates, 1999).

In Chapter 2, the concept of policy transfer is introduced as the major theoretical underpinning of this volume. While (semi)directed processes of transferring knowledge are central to policy transfer, this chapter focuses on the concept of policy translation that was introduced to cover processes associated with the 'travel of ideas' (Mukhtarov, 2009). Policy translation implies that new rules and practices are not simply applied, but actively reinterpreted before they are put into practice in another setting (Fadeeva, 2005). We analyse how this translation is taking place from a discourse analysis perspective. In doing so, we focus on how global ideas about adaptation were translated to the national and the local governance level in the Netherlands. The following research question is defined for the chapter as: *How has adaptation been interpreted in the Dutch water governance context and how have contextual factors influenced the local translation of these interpretations?*

Without its dunes and dykes, two-thirds of the Netherlands would flood at high sea and river levels (Huisman, 2004). Because of this, the Netherlands is projected to be one of the most vulnerable parts of Europe (EEA, 2006). Climate change manifests here as an increase in the occurrence and intensity of flooding risk, urban heating and local flooding from extreme showers. Recent research confirms that weather patterns in the Netherlands are currently changing (Van Dorland *et al.*, 2010).

The chapter first briefly presents our research methodology and the Dutch water management setting to provide a sketch of the study's case background. It then describes how adaptation has been interpreted in the Dutch water governance context by analysing the relevant storylines on adaptation. Following this, the focus is shifted towards the local level while analysing how local factors have influenced the local translation of adaptation and how the storylines have 'landed' locally. This chapter concludes with comments on the nature of the translation of the adaptation storylines as well as the most influential factors which explain the variation of local adaptation policies.

12.2 Methodology

The chapter draws on the Netherlands as a study area. First, we explore how adaptation was translated in storylines at the national level. Second, we analyse how these national storylines are translated at the local level. Our unit of analysis is the municipality as an administrative body consisting of civil servants that prepare and implement local policies. Local policymaking is the responsibility of the local alderman who is an elected governor for a period of four years.

For the analysis of the local adaptation process, we build upon the experiences of fourteen municipalities.[1] In selecting these cases, municipalities were identified where adaptation efforts were most likely and municipalities in which such efforts were most unlikely (George and Bennett, 2005). We hypothesized that variation in adaptation policies between municipalities could stem from differences in (1) degree of urbanization, (2) flooding experience and (3) an increasing flooding risk induced by climate change. First, regions were identified that are at an increasing risk of flooding using the National Flooding Map (Min. V&W, 2009, p. 67) and that either had or had not experienced flooding events in the recent past based on the International Disaster Database (CRED, n.d.). Within these regions both the smallest and largest municipality (in terms of population) were selected as study objects. The resulting selection included eight rural and six urban municipalities. Of the fourteen cases, eight are facing an increased risk of flooding, whereas four have experienced a flooding disaster.

The study draws on discourse analysis. Uniting an imaginary community under a shared world view, a discourse consists of people and a topic and defines meanings and solutions of problems related to the topic (Dryzek, 2005). Discourse analysis can be placed in the constructionist tradition assuming the existence of multiple socially created realities, instead of one governed by natural laws. This brings the analysis of meaning to the forefront: 'for interpretative environmental policy research, it is not an environmental phenomenon in itself that is important, but the way in which society makes sense of this phenomenon' (Hajer and Versteeg, 2005, p. 176). In his efforts on theorizing the discourse approach, Hajer (1995) introduced the storyline as an interpretation of a specific topic within the broader discourse.

In 'doing' discourse analysis, Yanow (2000) instructs the researcher to determine the actors and communities involved so as to distinguish the relevant discourses and their storylines. This is done by chronologically analysing the relevant storylines in the water management discourse that, as conversation streams, are discussing problems and solutions towards it. It is further seen that policy documents are being 'discharged' from a discourse at regular intervals reflecting the current state of affairs in terms of storylines. In analysing these storylines, we understand adaptation as the explicit reproduction of the term in policy documents (or in interviews) or more implicit as the consideration of climate change impacts in the formulation and execution of policy.

It has been demonstrated earlier that the local agenda for sustainable development was largely set by national governmental initiatives (Coenen, 2008). It has been argued that the adaptation concept was first defined nationally before it was brought to the lower levels of government (Klostermann *et al.*, 2009). Hence, first the contents of national policy documents on water management in general and adaptation in particular have been examined so as to identify the national perception of adaptation and how the local level was addressed in this. After this, interviews were held with local civil servants to verify their views and experiences on the issue. The in-depth interviews were conducted based on an open-ended questionnaire with civil servants responsible for policy areas such as spatial planning, water management and the environment. The studies were carried out in two rounds of which the first round of nine cases was a nationwide study, while the second round of five municipalities took place within the borders of one province.

12.3 The Netherlands and water management

Water, and in particular managing it, is strongly connected to the cultural context of the Netherlands. The Netherlands has developed a risk culture where effective coping mechanisms have been developed to limit losses from flooding (Mauch and Pfister, 2009). Van Dam (2010) introduces the concept of the amphibious culture to characterize the long and intensive tradition of the Dutch relationship of living with water. Examples of this tradition include an effective flood protection system (Underdal, 2010), a strong tradition of water-based transportation and traditional living on elevated grounds (Van Dam, 2010).

Another coping mechanism was the institutionalization of water management (Van Koningsveld *et al.*, 2008). Rather early in history, local and regional water boards were developed as a democratic government strictly responsible for the management of the water quantity and quality within their region, including flood risk prevention from regional water. Additionally, in 1798 Rijkswaterstaat was established as the national water board to pursue the national interests in water management (Tol and Langen, 2000).

Today, the national government protects nationally important waterways, large lakes and the coast, whereas the twenty-five water boards in the country secure the regional and local water bodies. Sufficient coastal protection is crucial due to a third of the national economy being at risk from flooding concerns (Van Koningsveld *et al.*, 2008). Lower in the hierarchy, at the regional level the province secures regional water interests, whereas municipalities are responsible for local precipitation drainage, sewage and groundwater management in urban areas. Table 12.1 includes an overview of the key ministries and associations presented throughout the paper and presents a brief description of their role in the adaptation policy process. The roles of the various ministries are currently changing due to recent mergers of the relevant ministries.

From a hydraulic perspective, the country was 'finished' in the course of the twentieth century with the realization of the Afsluitdijk ('Closure Dam') and the Delta Works (Disco, 1998, p. 55). The Afsluitdijk has closed off the Zuider Zee, whereas the Delta Works – constructed after the latest flood disaster in 1953 to prevent such a future flooding once and for all – mainly included a drastic shortening of the Dutch coastline by closing off the main

Table 12.1 Key ministries and associations involved in the adaptation policy process

Dutch abbreviation	*Full name*	*Role in the adaptation process*
V&W	Ministry of Transport, Public Works and Water Management	Emphasizing the water and safety aspects of adaptation. Has become the dominant actor in the adaptation process. Merged into Ministry of Infrastructure and the Environment as of 2010
VROM	Ministry of Housing, Spatial Planning and the Environment	Emphasizing the wider dimensions of adaptation outside of the water scope. Its voice has become less profound since its merge into Ministry of Infrastructure and the Environment as of 2010
LNV	Ministry of Agriculture, Nature and Food Quality	Limited involvement since its merged into Ministry of Economic Affairs, Agriculture and Innovation as of 2010
EZ	Ministry of Economic Affairs	Limited involvement since its merged into Ministry of Economic Affairs, Agriculture and Innovation as of 2010
IPO	Association of the Provinces of the Netherlands	Representing the regional level and co-author of the National Adaptation Strategy. Modest role
UvW	Association of Water Boards	Representing the Regional Water Authorities and co-author of the National Adaptation Strategy. Active role in emphasizing the water aspects of climate change and adaptation
VNG	Association of Dutch Municipalities	Representing the local level and co-author of the National Adaptation Strategy. Emphasizing the (difficulties in) implementing adaptation

river estuaries. However, these measures did not take the currently projected sea level rise into account, nor were they designed to cope with increasing rainfall or higher river discharges. Therefore, the Dutch are challenged to design and realize new large-scale projects to 'keep their feet dry' (Van Koningsveld *et al.*, 2008).

12.4 Storylines on adaptation in the Dutch context

Over time, three storylines emerged around the idea of adaptation: (1) a water accommodation storyline, (2) a climate proof storyline and (3) a safe delta storyline. These storylines are elaborated below. Table 12.2 presents an overview of the policy documents that are discussed in this section.

Table 12.2 Overview of policy documents presented

Period	*Name*	*Contents*	*Authors/participants*[1]
2000	Water Policy for the 21st century	Position paper on water accommodating policy including IWRM and adaptation	Ministry V&W
2003	National Agreement on Water	Agreement to jointly prepare the water system for the twenty-first century including adaptation	Ministry V&W, provinces (IPO), municipalities (VNG), water boards (UvW)
2007	Updated National Agreement on Water	Implementation plan of the previous National Agreement on Water	Ministry V&W, provinces (IPO), municipalities (VNG), water boards (UvW)
2009	National Water Plan	This national policy plan (2009–2015) describes measures to keep the country safe from flood risk, to have sufficient and clean water in the future and to make use of the (economic) water-related opportunities. It includes River Basin Management Plans as required by the European Water Framework Directive	Ministry VROM, Ministry V&W, Ministry LNV
2007	Make Space for Climate! National Adaptation Strategy	This national strategy targets to climate proof the country. In the long term, the aim is to embed adaptation in awareness, policy and regulation. In the short term, the aim is to initiate an impulse for climate adaptation measures	Ministry VROM, Ministry V&W, Ministry LNV, Ministry EZ, provinces (IPO), municipalities (VNG), water boards (UvW)
2007	Climate Agreement State–municipalities	Joint agreement on measures to prepare for climate change impacts. It includes inventorying local 'hooks' for adaptation	State, municipalities (VNG)
2009	Delta programme	This programme aims to protect the Netherlands against flooding and to ensure supplies of freshwater in the future. Areas related to the plan include the environment and the economy, nature, agriculture and recreation. Implementation of the programme is directed by the Delta Commissioner whose position is anchored in the Delta Act	Ministry V&W, Ministry LNV, Ministry VROM

Note:

1 See Table 12.1 for clarification of the abbreviations

The water accommodation storyline

Dutch water managers embraced the adaptation concept in the beginning of the twenty-first century. In the preceding period, attention for Integrated Water Resources Management (IWRM) had been on the rise since the mid-1980s (Mostert, 2006; Wesselink, 2007). Following large-scale flooding events in the 1990s,[2] questions arose in the governmental and academic discourse as to whether traditional technological interventions would be sufficient to cope with future impacts (Voogd, 2006). The water management discourse then became more holistic and collectively addressed the increasingly manifesting complex and interdependent problems of climate change impacts, increasing urbanization and soil subsidence (UvW, 2002).

The new water governance approach covering both IWRM and water-related adaptation was formally introduced in a national position paper in 2000 (Min. V&W, 2000). It stated that a new approach of proactive (e.g. adaptive) governance was preferred over the traditional reactive response. The emphasis of anticipating future developments included the principle of 'accommodating water', which soon became the general storyline in the water management discourse (Wiering and Immink, 2006).[3] Inextricably bound to the new approach, adaptation was quickly integrated into the discourse.

The storyline interprets climate change in general as increasing extremes in precipitation that result in enlarging (local) flooding risks. Hence, 'solutions' include measures for water storage and adjusted discharge capacity. It is emphasized that measures should be integrated in terms of the parties involved, its geographical scope and mutual connections. However, the water accommodation story restricts its appeal to the major water authorities.

The storyline was brought into practice in water policies drawn up in the early years of the twenty-first century. In 2003, a governmental agreement was reached between the national, regional and local authorities on the new ambitions in water management (see Table 12.2). Aspiring to meet various complex water questions, it was agreed that the entire Dutch water system should be 'in good order' by 2015 in terms of water safety, water shortages and water quality – and that it is kept at that status from then on. The agreement was updated by 2006 to include a practical approach and it was agreed that the new approach would be incorporated into existing policies, such as the urban local zoning plan and regional planning. The first National Water Plan in 2009 formalized the previously agreed arrangements (Min. V&W, 2009).

The new national water strategy has considerable local effects as it prescribes that municipalities carry out the 'municipal water task', which is the obligation for municipalities to adjust their urban water system so it can cope with future climatic changes. This implies technical measures such as additional water storage, smart discharging, and decoupling rainwater discharge from the sewer system. In order to reach their goals, all governmental layers agreed to use their respective powers – and to finance the measures themselves (Min. V&W, 2009).

The climate proof storyline

Global climate negotiations provide a framework for the development of national adaptation policies. Adaptation has increasingly gained international attention since 2001 (Schipper, 2006). At the Conference of the Parties (COP) of the UNFCCC in 2002, adaptation was stated as being of high priority to all countries (Adger *et al.*, 2005). In addition, at the latest COP, in 2011, it was decided that 'planning for adaptation at the national level is a continuous, progressive and iterative process, the implementation of which should be based on nationally identified priorities, including those reflected in the relevant national documents' (UNFCCC, 2011, p. 80).

For the Netherlands, it was not until 2006[4] that adaptation was given broader policy attention outside of the water governance discourse discussed above. Inspired by a senator's appeal in 2005, the country's first long-term vision on spatial planning was developed that would also include climate change impacts. As a result of multiple government efforts, the National Adaptation Strategy (NAS) was drawn up in 2007 by four ministries and the regional and local governments. The new storyline strongly relates adaptation to spatial planning and covers the country as a whole (Min. VROM, 2007).

The climate proof storyline has a much broader view on climate change and adaptation than the water accommodation storyline has. Climate change is interpreted in terms of scenarios including a wide variety of climatic and weather changes. This is directly linked to societal impacts regarding safety, the environment, biodiversity and the economy. Solutions are sought in the direction of spatial measures, together with societal and governmental efforts. The storyline aims for spatial adaptation that would result in the 'climate-proofing' of the Netherlands. In order to reach this goal, all levels of government are expected to contribute.

The relevance of the local level was formalized in a 'hotspot' approach. This approach identifies geographical areas of relative high vulnerability and analyses these locations in the two leading national research programmes on climate change (Climate Changes Spatial Planning from 2004–2011, followed by Knowledge for Climate from 2007–2012). While the first programme researched adaptation options for three local sites, the second programme turned the hotspot approach into its backbone and has been investigating eight regional and local hotspot areas. Acting as 'experimental gardens' for the application of adaptation in particular geographic conditions, the hotspots acted as bottom-up studies on areas with particular geographic, economic or natural conditions.

Despite its elaborated scope and options for responding, progress on the NAS was impeded due to political competition over the adaptation issue between the Ministry of Housing, Spatial Planning and the Environment and the Ministry of Transport, Public Works and Water Management (Biesbroek *et al.*, 2010). As a result, the climate proof storyline was soon 'taken over' by a new storyline as expressed in the Delta programme.

The safe delta storyline

Initiated by one of the leading water management authorities, the Ministry of Transport, Public Works and Water Management presented the Delta programme in 2010. This extensive national adaptation programme represents the third storyline, namely that of 'the safe delta'. It associates climate change with natural hazards such as the North Sea flood of 1953 that took 1,800 lives. The storyline argues that the hazardous impacts of climate change, particularly increasing flooding risks and shortages of freshwater provisions, require drastic measures to prevent disasters in the future. Flood protection, interpreted as being of vital importance in the safe delta storyline, is one of the key 'solutions' to keep the Netherlands safe for future generations (Min. V&W, 2010). Compared to the previous storylines that emphasized the need for multi-governmental efforts, the national government has now taken the lead in directing the adaptation process. Framing adaptation as a national priority in terms of flood protection legitimizes a strong role of the national government.

Some of the climate proof storyline still remains in the safe delta storyline, as spatial planning is the subject of one of the sub-programmes, 'New construction and restructuring'. This sub-programme 'explores how spatial measures can contribute to limiting the consequences from flooding, extreme rain, a longer period of drought and extreme heat' (own translation, taken from Delta Commissioner, n.d.). The programme, run by the relatively new Ministry of Infrastructure and the Environment, is currently preparing tools for regional and local planners to improve the consideration of adaptation in new projects that involve restructuring the build-up environment. Given the recent emergence of the safe delta storyline, a full analysis of the discourse surrounding it is not yet possible and only preliminary results can be used in the remainder of this chapter.

12.5 The role of contextual factors in translating adaptation at the local level

The previous sections have introduced how adaptation has been interpreted by national authorities. Choosing now to focus on the local level, three types of contextual factors are hypothesized to show the variation present in the translation of adaptation: (1) level of urbanization, (2) flooding experience and (3) increased risk from flooding.

The role of contextual factors

The degree of urbanization shows physical and organizational variation between the urban and rural cases. In terms of physical variation, the impacts from climate change will have lesser acute consequences in rural municipalities than in the case of urban areas. When necessary, excess run-off can be discharged outside of inhabited areas (interview at Wierden municipality). In urban cases, as a result of their paved character, cities are increasingly confronted with local flooding caused by extreme rainfall.

Second, the degree of urbanization also provides for major differences of an organizational dimension. Rural municipalities were observed to have a narrower focus on their duties than their urban counterparts. Rural case areas have less room for specialization and exploration of new topics, but show a greater capacity for quick actions as their procedural processes take less time. On the contrary, urban case areas have the advantage of having more expert staff available to be aware of and address issues that are beyond the local scope. This results in them undertaking more front-running activities, such as developing more general policies on adaptation. In these cases, respondents indicated that they perceive adaptation as a broadening of the more generally accepted mitigation policy towards climate change (van den Berg and Coenen, 2012). However, urban policymaking also involves more complex internal procedures of consulting and negotiating to integrate the generally more sophisticated urban policies into a coherent approach.

In the case of experience with flooding, two factors were found to explain variation between experienced versus non-experienced municipalities. First, experience contributes to increased awareness. Having had a flooding experience enables civil servants to better visualize – and hence convince elected officials – the possibility of future floods. However, this does not automatically turn into action in terms of policymaking. Particularly in the rural cases, it appears that the experienced actors do not see a need to take action themselves, but rather trust the national government to provide the necessary protection.

Second, flooding events often results in improved physical and institutional infrastructure for flood protection. In the aftermath of a serious flood, the national government usually supports physical measures to prevent future damage (for example through fortification of dykes and dams such as in the cases of Nijmegen and Kampen after the flooding events in the 1990s). These experiences can also function as a window of opportunity within the municipal organization as is shown in the case of Nijmegen where a Water Plan was formulated to improve the entire water management system of the city (interview Nijmegen). Following from the Nijmegen example, this policy document has now become more common in other municipalities that experience local flooding.

As for the factor of increased flooding risk, the flood-prone municipalities have set up a disaster management structure that comes with the flood-prone status. Because of their flood-prone status, these municipalities have direct contacts with the province where in general a higher level of concern for adaptation is present. More generally, increasing variability in daily weather patterns is relevant for all municipalities. In their current disaster plans, municipalities however indicate that extreme weather events are not being prepared for since these events are not being anticipated (van den Berg and Coenen, 2012). Physical adaptations are taking place, also within the municipal water task, as everywhere water accommodation is being increased through sewer and run-off adjustments. Locally, however, no adaptation measures outside of the scope of the water management discourse have been encountered.

Translating the storylines at the local level

And how have the three storylines been translated at the local level? The water accommodation storyline obtained a stable position as all of the studied municipalities are now implementing measures to enlarge the accommodation of water. Through this storyline adaptation in the form of the municipal water task (introduced above) was brought to the local level. This obligation could often directly be connected to the local setting that is increasingly facing local flooding events from extreme showers (RIONED, 2007).

Within the climate proof storyline municipalities are committed to inventorying possibilities for adaptation in the areas of spatial planning, water management and public health. The case studies show two such local inventorying processes for ongoing adaptation projects (Breda and Utrecht). Another manifestation of the climate proof storyline was the introduction of the hotspot approach where local and regional cases have been selected to research adaptation options for the area. However, while most of the municipalities of our study were participating in the programme these municipalities often prove not to be familiar with the storyline – or they feel excluded from it (Nijmegen).

Only the city of Utrecht is part of a hotspot, and the level of adaptation measures here is found to be rather high. The city has, together with the province, established the 'Climate adaptation workshop' where civil servants from various governments and policy domains jointly design a new residential district under the hotspot approach of the Knowledge for Climate research programme. The primary explanations for this high level of action are understood as related to factors outside of our case selection criteria (e.g. political prioritization and the city's position as the fourth city of the country resulting in close national contacts).

As indicated above, the safe delta storyline is rather recent and is therefore only very minimally included in our empirical data. This storyline will probably reach more municipalities than the climate proof storyline as its nationwide approach towards spatial adaptation has a more direct and problem-solving approach than the previous initiatives. However, with its primary focus on water it is unlikely that this storyline will generate a more robust translation of adaptation outside of the 'hard' domains of water and spatial planning.

12.6 Conclusions

This chapter addresses how the global idea of climate change adaptation is being translated in the Netherlands at the national and the local level. We distinguished several lines of policy that were recently initiated to implement national ambitions in terms of water management and climate-proofing. Adaptation was introduced as a water governance issue in the early years of 2000 as part of the water accommodation storyline. Expanding the water focus on adaptation with a spatial orientation, the climate proof storyline delivered a National Adaptation Strategy by 2007. Currently, adaptation ambitions have gone beyond increasing the safety level of the Netherlands with respect to flooding and ensuring sufficient fresh water for future generations within the safe delta storyline through the delivery of the Delta programme in 2009.

To understand how contextual factors influence the translation of policy ideas, we selected fourteen Dutch municipalities as case studies. The empirical data shows that in terms of urbanization, cities have an advantage over their rural counterparts as managing a city requires more specialist staff and the impacts on cities are more acute – and so are the measures. In case of flooding experience, a higher level of awareness and a physical and institutional flood protection system restrict future flooding. However, it is the national level that is envisioned as being responsible for the continuation of advanced flood protection. In the case of increasing flooding risk, their flood-prone status comes with multilevel contacts and a distinct disaster plan, but adaptation has not been introduced here.

The water accommodation storyline has obtained a clear footing, as it has become the local standard for municipal water management. Targeting at an increase of the resilience of local and regional water systems, adaptation has become an intrinsic component of water management. Manifestations of the climate-proofing storyline were only sporadically encountered, while the recent safe delta storyline is unfortunately too recent to be covered in our data. Having a rather direct character, however, this storyline is seen to have an increased chance of actually resulting in local adaptation policymaking outside of the dominant water management domain.

The amphibious culture of the Netherlands was introduced to illustrate the special relationship between the Dutch and water, which was assumed to be an important element in the water focus seen in Dutch adaptation discourse. However, studies on other EU contexts also report of a strong water focus in adaptation (e.g. Tompkins *et al.*, 2010; Swart *et al.*, 2009). Being a more general mechanism, a focus on water-related adaptation is reasonably connected with observations from the case studies. There is a strong need at the local level for visualizing climate change impacts – which can be done most effectively for many of the water-related impacts, e.g. intense rainfall or riverine flooding. Other impacts such as an increase in extreme weather events in general, or temperature rise, are harder to imagine and, accordingly, so is preparing for it: local contingency plans indicate that extreme weather is not foreseen – hence, preparing for it rarely occurs.

Acknowledgements

Research for this study was funded by NWO, Waterschap Groot Salland and the Norwegian Research Council. The author would like to thank the editors for their feedback on earlier versions of the paper.

Notes

1 The municipalities include (alphabetically): Almere, Breda, Enschede, Kampen, Millingen aan de Rijn, Nijmegen, Noord-Beveland, Rijssen-Holten, Schiermonnikoog, Terschelling, Tubbergen, Utrecht, Wierden and Zwolle.

2 In 1995, 250,000 people and even more livestock had to be evacuated from the Rivierengebied ('Rivers Area') when water levels of the Rhine and Meuse rivers rapidly rose threatening the dykes of the area.

3 Many activities for the EU Water Framework Directive correspond to the new water governance approach, e.g. aiming at sufficient supply of high quality surface water and groundwater as needed for sustainable water use (European Communities, 2000).

4 Prior to this, climate change research has been facilitated in the Climate Changes Spatial Planning (2004–2011) research programme 'feeding' policy making for the NAS. This programme was supported by the Dutch Government from the Economic Structure Enhancing Fund (FES).

References

Adger, W.N. (2001) 'Scales of governance and environmental justice for adaptation and mitigation of climate change', *Journal of International Development*, vol. 13, no. 7, pp. 921–931.

Adger, W.N., Arnell, N.W. and Tompkins, E.L. (2005) 'Successful adaptation to climate change across scales', *Global Environmental Change Part A*, vol. 15, no. 2, pp. 77–86.

Biesbroek, G.R., Swart, R.J., Carter, T.R., Cowan, C., Henrichs, T., Mela, H., Morecroft, M.D. and Rey, D. (2010) 'Europe adapts to climate change: Comparing national adaptation strategies', *Global Environmental Change*, vol. 20, no. 3, pp. 440–450.

Coenen, F. (2008) 'The local level: New interpretations of governance for sustainable development in the Netherlands', in S. Baker and K. Eckerberg (eds) *In Pursuit of Sustainable Development: New Governance Practices at the Sub-national Level in Europe*, Routledge, Abingdon.

CRED (Centre for Research on the Epidemiology of Disasters) (n.d.) *International Disaster Database*, www.emdat.be.

Cutter, S.L. (2003) 'The vulnerability of science and the science of vulnerability', *Annals of the Association of American Geographers*, vol. 93, no. 1, pp. 1–12.

Delta Commissioner (n.d.) *Nieuwbouw en herstructurering* [New urban developments and restructuring], online, available at: www.deltacommissaris.nl/organisatie/deelprogrammas/nieuwbouw_en_herstructurering, accessed 9 November 2012.

Disco, C. (ed.) (1998) 'Techniek in ontwikkeling, waterstaat, kantoor en informatietechnologie [Technology in development, water management, office and information technology], *Techniek in Nederland in Twintigste Eeuw*, vol. I, Walburg Pers, Zutphen, Netherlands.

Dryzek, J.S. (2005) *The Politics of the Earth: Environmental Discourses*, 2nd edition, Oxford University Press, Oxford/New York.

EEA (European Environment Agency) (2006) *Vulnerability and Adaptation to Climate Change in Europe*, European Environment Agency, Copenhagen.

European Communities (2000) *Directive 2000/60/EC of the European Parliament and of the Council establishing a framework for the Community action in the field of water policy*, online, available at: http://eur-lex.europa.eu/LexUriServ/LexUriServ.do?uri=CELEX:32000L0060:EN:NOT, accessed 15 October 2012.

Fadeeva, Z. (2005) 'Translation of sustainability ideas in tourism networks: Some roles of cross-sectoral networks in change towards sustainable development', *Journal of Cleaner Production*, vol. 13, no. 2, pp. 175–189.

George, A.L. and Bennett, A. (2005) *Case Studies and Theory Development in the Social Sciences*, MIT Press, Cambridge, MA.

Hajer, M.A. (1995) *The Politics of Environmental Discourse: Ecological Modernization and the Policy Process*, Oxford University Press, Oxford/New York.

Hajer, M. and Versteeg, W. (2005) 'A decade of discourse analysis of environmental politics: Achievements, challenges, perspectives', *Journal of Environmental Policy and Planning*, vol. 7, no. 3, pp. 175–184.

Huisman, P. (ed.) (2004) *Water in the Netherlands*, revised edition, Netherlands Hydrological Society (NHV), Delft.

Klostermann, J.E.M., Biesbroek, G.R. and Gupta, J. (2009) 'Multilevel klimaatbeleid in Nederland: Mitigatie en adaptatie' [Multilevel climate policy in the Netherlands: mitigation and adaptation], *Bestuurskunde*, vol. 18, no. 4, pp. 17–26.

Mauch, C. and Pfister, C. (eds) (2009) *Natural Disasters, Cultural Responses: Case Studies Toward a Global Environmental History*, Lexington Books, Lanham, MD.

Min. V&W (Ministry of Transport, Public Works and Water Management) (2000) *Anders omgaan met water. Waterbeleid voor de 21 eeuw* [Handling water differently. Water policy for the 21st Century], Ministerie van Verkeer en Waterstaat, The Hague.

Min. V&W (Ministry of Transport, Public Works and Water Management) (2009) *Nationaal Waterplan* [National Water Plan], The Hague, Ministerie van Verkeer en Waterstaat, Ministerie van Landbouw, Natuur en Voedselkwaliteit and Ministerie van Volkshuisvesting, Ruimtelijke Ordening en Milieubeheer.

Min. V&W (Ministry of Transport, Public Works and Water Management) (2010) *Deltaprogramma 2011: Werk aan de delta. Investeren in een veilig en aantrekkelijk Nederland, nu en morgen* [Delta Programme 2011: Working on the delta. Investing in a safe and attractive Netherlands, now and tomorrow], The Hague, Ministerie van Verkeer en Waterstaat, Ministerie van Landbouw, Natuur en Voedselkwaliteit and Ministerie van Volkshuisvesting, Ruimtelijke Ordening en Milieubeheer.

Min. VROM (2007) *Maak ruimte voor klimaat! Nationale adaptatiestrategie – de beleidsnotitie* [Make space for climate! National adaptation strategy – the policy paper], The Hague, Ministerie van Volkshuisvesting, Ruimtelijke Ordening en Milieubeheer, Ministerie van Verkeer en Waterstaat, Ministerie van Landbouw, Natuur en Voedselkwaliteit, Ministerie van Economische Zaken, Interprovinciaal Overleg, Vereniging van Nederlandse Gemeenten and Unie van Waterschappen.

Mostert, E. (2006) 'Integrated water resources management in the Netherlands: How concepts function', *Journal of Contemporary Water Research and Education*, vol. 135, no. 1, pp. 19–27.

Mukhtarov, F. (2009) 'The hegemony of integrated water resources management: A study of policy translation in England, Turkey and Kazakhstan', PhD thesis, Central European University at Budapest, Hungary.

Nilsson, A.E., Gerger Swartling, Å. and Eckerberg, K. (2012) 'Knowledge for local climate change adaptation in Sweden: Challenges of multilevel governance', *Local Environment*, vol. 17, no. 6/7, pp. 751–767.

Parry, M.L., Canzani, O.F., Palutikof, J.P., van der Linden, P.J. and Hanson, C.E. (eds) (2007) *Climate Change 2007: Impacts, Adaptation and Vulnerability. Contribution of Working Group II to the Fourth Assessment Report of the Intergovernmental Panel on Climate Change*, Cambridge University Press, Cambridge.

RIONED (2007) *Klimaatverandering en hevige buien. Persbericht regenwateroverlast* [Climate change and heavy showers. Press release rain flooding], www.riool.net, accessed 5 September 2012.

Schipper, E.L.F. (2006) 'Conceptual history of adaptation in the UNFCCC process', *Review of European Community and International Environmental Law*, vol. 15, no. 1, pp. 82–92.

Smit, B. and Pilifosofa, O. (2001) 'Adaptation to Climate Change in the Context of Sustainable Development and Equity', in J.J. McCarty, O.F. Canzani, N.A. Leary, D.J. Dokken and K.S. White (eds) *Climate Change 2001: Impacts, Adaptation, and Vulnerability. Contribution of Working Group II to the Third Assessment Report of the Intergovernmental Panel on Climate Change*, Cambridge University Press, Cambridge, pp. 877–913.

Storbjörk, S. (2010) 'It takes more to get a ship to change course: Barriers for organizational learning and local climate adaptation in Sweden', *Journal of Environmental Policy and Planning*, vol. 12, no. 3, pp. 235–254.

Swart, R.J., Biesbroek, G.R., Binnerup, S., Carter, T.R., Cowan, C., Henrichs, T., Loquen, S., Mela, H., Morecroft, M., Reese, M. and Rey, D. (2009) *Europe Adapts to Climate Change: Comparing National Adaptation Strategies, Partnership for European Environmental Research* (PEER), Helsinki.

Tol, R.S.J. and Langen, A. (2000) 'A concise history of Dutch river floods', *Climatic Change*, vol. 46, no. 3, pp. 357–369.

Tompkins, E.L., Adger, W.N., Boyd, E., Nicholson-Cole, S., Weatherhead, K. and Arnell, N. (2010) 'Observed adaptation to climate change: UK evidence of transition to a well-adapting society', *Global Environmental Change*, vol. 20, no. 4, pp. 627–635.

Underdal, A. (2010) 'Complexity and challenges of long-term environmental governance', *Global Environmental Change*, vol. 20, no. 3, pp. 386–393.

UNFCCC (2011) *Report of the Conference of the Parties on its seventeenth session*, held in Durban from 28 November to 11 December 2011, FCCC/CP/2011/9/Add.1, online, available at: http://unfccc.int/meetings/durban_nov_2011/session/6294.php, accessed 30 September 2012.

UvW [Union of Water Boards] (2002) '*Waterbeheer 21e eeuw: aanleiding, afspraken en maatregelen* [Water 21st century: rise, agreements and measures], The Hague, Unie van Waterschappen [Union of Water Boards].

Van Dam, P. (2010) *De amfibische cultuur. Een visie op watersnoodrampen* [The amphibious culture. A vision on flooding disasters], inaugural speech for the position of endowed professor of water management history, Free University of Amsterdam, Amsterdam.

van den Berg, M. and Coenen, F. (2012) 'Integrating climate change adaptation into Dutch local policies and the role of contextual factors', *Local Environment*, vol. 17, no. 4, pp. 441–460.

Van Dorland, R., Dubelaar-Versluis, W. and Jansen, B. (2010) *De staat van het klimaat 2009. Actueel onderzoek en beleid nader verklaard* [The state of the climate 2009. Recent research and policy further explained], Platform Communication on Climate Change (PCCC), De Bilt/Wageningen, Netherlands.

Van Koningsveld, M., Mulder, J.P.M., Stive, M.J.F., Van der Valk, L. and Van der Weck, A.W. (2008) 'Living with sea-level rise and climate change: A case study of the Netherlands', *Journal of Coastal Research*, vol. 24, no. 2, pp. 367–379.

Voogd, H. (2006) 'Combating flooding by planning: Some Dutch experiences', DISP-ZURICH, vol. 164, no. 1, pp. 50–58.

Wesselink, A.J. (2007) 'Integraal waterbeheer: de verweving van expertise en belangen', PhD thesis, University of Twente, Enschede.

Wiering, M. and Immink, I. (2006) 'When water management meets spatial planning: A policy-arrangements perspective', *Environment and Planning C: Government and Policy*, vol. 24, no. 3, pp. 423–438.

Wilbanks, T.J. and Kates, R.W. (1999) 'Global change in local places: How scale matters', *Climatic Change*, vol. 43, no. 3, pp. 601–628.

Wilson, E. (2006) 'Adapting to climate change at the local level: The spatial planning response', *Local Environment*, vol. 11, no. 6, pp. 609–625.

Yanow, D. (2000) *Conducting Interpretive Policy Analysis*, Qualitative research methods series 47, SAGE, Thousand Oaks, CA.

13 Adaptive responses to water scarcity

Transfer of governance approaches across South and North Europe

Stefan Kuks and Cheryl de Boer

13.1 Introduction

Climate change requires that societies rethink their prevailing approaches to water management. One impulse for this is that they are increasingly confronted with changing rainfall patterns. This results not only in difficulties with flooding (excessive rainfall over a short period of time), but also with drought (not enough rainfall over a long period of time). The latter of the two types of water issues relates to water scarcity problems that have long been experienced in the more arid Mediterranean countries within Europe. Natural conditions of the water basin play an important role in determining the frequency and amount of precipitation and storage in an area as well as the changing climate and rainfall patterns. As such, a number of issues related to water scarcity are also becoming important to many Northern European countries. This chapter discusses the contextual conditions that are of importance in the development of adaptive policy responses to these water scarcity issues.

While differences of opinion exist with respect to the exact definition of water scarcity, it is often dealt with as a situation where the availability of water does not meet the demands of people and nature (White, 2012). To reduce the effects of water scarcity, a policy approach should focus on socio-economic factors related to the water utilization pattern. Adaptive responses should reconsider both traditional and more recently recognized natural and recreational water demands and take into account rival demands for rather vulnerable use functions in society. Traditional demands include mainly economic functions (e.g. guaranteeing the navigability of rivers, guaranteeing the use of rivers for cooling energy power stations, preventing saltwater intrusion into aquifers for drinking water and agricultural crop growing). Adaptive responses are crucial for managing these functions as well as for the more natural and recreational values of water systems. The result is often the development of minimum flow requirements.

The results of the European Commission supported Aquadapt project (2000–2005) serve as the empirical basis upon which this chapter is based

(Kuks, 2005a, 2005b). The project team consisted of thirteen academic and industrial partners from seven countries and contributed to the knowledge base available for those involved in strategic planning and management of water resources in semi-arid environments (Aquadapt, 2003). The changing supply and demand patterns were an important driver for this project. Four cases are presented in this chapter based on the data collected by country teams in Alicante for the Spanish case, in Montpellier for the French case, in Cranfield for the United Kingdom (UK) case, and in Twente for the Dutch case. The project teams investigated the way in which governmental actors deal with and respond to problems of drought and water deficiency in an adaptive way. The two cases in Southern Europe (the Marina Baja catchment area in Spain and the Hérault river basin in France) were compared with two cases in Northern Europe (the Nene river basin in the UK and the Maas river basin in the Netherlands). The Aquadapt research was focused on identifying the changes in water utilization patterns and water policy responses to those changes in each of the chosen countries. Researchers in the four countries looked at the policy approaches that were adopted and implemented, as well as at the underlying governance regimes (the institutional setting) from which the policy approaches resulted in order to assess the prevailing governance regime. The aim of Aquadapt was to formulate general conclusions on how to make governance regimes more adaptive towards water scarcity. This chapter recognizes that adaptive policy responses are a key aspect required for water governance (de Boer *et al.*, 2011) and connects the presence of these responses to contextual conditions (outlined in the following section) that are important to achieve regime change (institutional change) and policy change (Bressers and Kuks, 2008). The next section also summarizes the theoretical framework that has been used for the comparative analysis of the case studies. Following this, we analyse and discuss the case studies from Spain, France, the UK and the Netherlands respectively. The final section provides conclusions about which contextual factors were observed to influence the emergence of adaptive policy responses in the four cases and how this relates to institutional change in water governance regimes.

12.2 Framework for comparative analysis

The four cases in South and North Europe focus on adaptive responses to changes in the water utilization pattern. Each country is investigated at the level of a specific case, while also taking into account the influence of contextual factors from outside the case area. First, we discern changes or trends in the utilization pattern during a period of three decades (1970–2000). Second, we identify adaptive responses during this period by considering the governance regime at the case level. We use the five dimensions of the governance regime as described in Chapter 3 of this book, and for each of these five dimensions, we investigate whether adaptive or reactive responses are present. Third, we relate these responses to changes in the utilization pattern or other contextual factors. Finally, we interpret the meaning of adaptive responses in terms of sustainability and discuss ways to enhance the future adaptive capacity of water governance regimes in coping with water scarcity.

Water utilization patterns resulting in sense of urgency

Water utilization patterns are influenced by both natural factors that determine the resource status (water availability and chances for water utilization), and socio-economic factors that determine the resource use. Natural factors refer to the natural conditions of a water basin, as well as conditions developing related to climate change and changing rainfall patterns. Socio-economic factors refer to traditional water demands that may be expanding and creating homogeneous rivalries (users demanding similar services from a limited stock), new water demands or values (like natural, ecological or recreational values of water) resulting in heterogeneous rivalries (between different use types) and demands for other natural resources, which also creates rivalries (such as land-use claims in floodplains). Natural factors and socio-economic factors can thus together create a certain sense of urgency in terms of water scarcity.

Governance regime (institutional setting)

In 2000, the European Union adopted the Water Framework Directive (WFD) to streamline the European Union myriad water legislation into one overarching strategy. The WFD imposes a general requirement for ecological protection and a minimum chemical standard for all surface waters. The key element of the directive is the introduction of a model for water management based on 'river basins', or geographical areas, rather than on administrative or political boundaries. The key aims of the WFD are set out in Directive 2000/60/EC (additional detail added by authors in following parentheses):

- expanding the scope of water protection to all waters, surface waters and groundwater;
- achieving 'good status' for all waters by a set deadline;
- water management based on river basins (following the example of e.g. Maas, Rhine and Scheldt across-the-borders river basin management);
- achieving a 'combined approach' of emission limit values and quality standards;
- getting the prices right (adequate water pricing acts as an incentive for the sustainable use of water resources); and
- getting the citizen involved more closely and streamlining legislation (replacing seven old water directives).

The above aims are seen to address the five dimensions of governance distinguished in Chapter 3 in the following ways:

1 levels and scales: water management at river basin scale;
2 actors: getting the citizen and all users involved more closely (participation);
3 perceptions and ambitions: expanding the scope and achieving 'good status' for all water bodies;
4 strategies and instruments: streamlining legislation and combined approach (integration); and
5 responsibilities and resources for implementation: getting the prices right (redistribution of costs).

The institutional setting in each case is identified and described according to these five dimensions.

Policy initiatives: reactive and adaptive responses

The following directions for institutional change are possible ways a governance regime can develop towards more institutional sustainability: (1) restructuring levels and scales (positioning the river basin level and organizing water management based on watershed boundaries); (2) involving new actors (developing participatory arrangements for involvement of all users and stakeholders with an interest); (3) reformulating the policy problem (developing an integral vision including all water values); (4) integrating policies (using integrated water legislation, integrated planning and integrated water resource management); (5) redistributing resources (limiting property and use rights, internalizing costs, full cost recovery) (Bressers *et al.*, 2004; Bressers and Kuks, 2005, 2006).

A policy initiative is understood in this chapter as a response in terms of a change to one or more dimensions of the governance regime. These initiatives can be both adaptive and reactive. An adaptive response is defined here as an active learning process in which a decision path has been adopted from a range of possible alternatives. Such a response is the outcome of a consultative process in which numerous institutions with different opinions and resources have collaborated (Pahl-Wostl *et al.*, 2007). A reactive response, on the other hand, is defined here as one that does not learn from past experience but attempts to foster solutions, which are rather short term or satisfy one section of the population or a type of institution. A reactive response is one in which alternative options were either not discussed or not feasible.

Other contextual (institutional-based) factors

In Chapter 3, various types of contextual factors and actor characteristics are distinguished. In this chapter, we specifically focus on the shared values, shared cognitions, and mutual dependencies among actors. In slight contrast to the terminology used in Chapter 3, we choose to focus on these 'shared' actor characteristics as being part of the context and hence treat them as contextual factors that behave as mechanisms that determine if the five dimensions of a governance system will change. In case of *shared values*, adaptation may arise from the tendency of actors to act from a set of constant and coherent values. More specifically, Bressers and Kuks (2004) found indications for the relevance of a 'tradition of cooperation' in a dominant policy ideology that supports integration. Actors that experienced positive examples of integration also exhibited mutual respect and trust in 'fair play'. In case of *shared cognitions*, adaptation arises from the tendency of actors to use a common reference frame to interpret cognitions. More specifically, Bressers and Kuks (2004) found indications for the relevance of 'joint problems and joint opportunities'. Under these circumstances common knowledge bases existed from respected sources on relevant problems and opportunities. Additionally information symmetry was found between the involved actors as well as a sense of responsibility for the future and respect for each other's interests. In case of *mutual dependencies* among actors, adaptation may arise from the dependence of actors on each other's resources. Bressers and Kuks (2004) also found indications for the relevance of 'institutional interfaces'. These interfaces determine the clarity of assigned responsibilities (to prevent territorial battles) and the legal or practical possibilities to protect negotiated compromises from continuous litigation. Actors, independent or within the administration, can also function as part of the institutional interface as brokers with solely process objectives. Bressers and Kuks (2004) also found indications for the relevance of a so-called 'credible alternative threat'. This can take the form of a sufficient imbalance of power favouring a dominant actor to enable unilateral action, the availability of information on alternative options to 'solve' the problem from the perspective of the dominant actor's perspective or an alternative option that has more severe consequences for the other stakeholders than the specific form of integration would have.

The following sections provide the empirical data from the four case studies. The above contextual factors are used to assess each of the governance regimes with respect to the presence and influence of these shared values.

13.3 Spain (Marina Baja catchment area)

The Marina Baja is a catchment area on Spain's eastern coast in the autonomous region of Valencia. Tourism along the coastline is very important for the Spanish economy, but also contributes to water stress. Fruit trees (especially orange and lemon trees), olive trees and almond trees are cultivated here and the population has significantly increased in the last twenty years.

Water utilization pattern

The Marina Baja area suffers from water scarcity and saltwater intrusion due to overexploitation. This is partly the result of *natural conditions.* It is a semi-arid area with a decreasing annual rainfall and frequent drought periods. A severe drought occurred in 1978. In addition, there are frequent floods in the autumn and the area generally experiences irregular rainfall patterns. The scarcity is also the result of *socio-economic conditions* since there is an overexploitation due to agriculture in inland areas (a decrease of rain-fed agriculture, an increase of irrigated agriculture), overexploitation due to urban growth and development and strong growing tourism in the coastal area.

Policy responses

Since the 1960s, there have been intra-basin transfers within the Jucar river basin district as well as the construction of new dams and wells in the Marina Baja area. Use limitations and recommendations have not been put in place to reduce consumption (demand management), although public bodies have been aware since the 1960s that urban growth and tourism would create water problems for the area. Since the creation of the 1978 Marina Baja Water Consortium, these more reactive responses to water scarcity have given way to more *adaptive responses.* This consortium is an arrangement between municipalities and farmers to reallocate water in such a way that treated wastewater from urban and tourist areas can be reused in agriculture. Municipalities in the coastal area now cover most of the investment and exploitation costs, which keeps the price of irrigated water as low as possible. The required contribution from the general budget is justified by the importance of tourism for the area's economy.

Governance regime

In terms of *levels and scales,* we observe that integrative coordination is missing between the Spanish administrative water management levels. The Environment Ministry supervises the river basin authorities, which develop water basin level policy. However, the Jucar river basin authority is isolated from the Autonomous Community Valencia and the municipalities, including the Marina Baja Water Consortium. The consortium is an initiative by municipalities, but is not under the supervision of the river basin authority. The WFD suggests concerns about the appropriate scale of intra- and inter-basin transfers with respect to sustainable river basins and the relevant basin boundaries. In terms of *actors and networks* we observe that the consortium is a closed community, protecting its strong economic interests. It is a corporative institution, not very open to rival interests (such as the interests of the ecosystem) or linking up with land-use development.

There is a low level of public participation and a lack of access to information. The consortium is locally embedded, but not in the top-down structure of the administration. In terms of *perspectives and objectives* we observe that an inconsistency in the National Hydrological Plan of 2001 on the one hand extended the eco-perspective but on the other hand was encouraging an increase in the number of dams and inter-basin transfers. Since 2004, the central government has developed a policy approach that aims more at addressing resource constraints and solutions in terms of desalinization and demand management. However, at regional and local levels the resource is not considered to be a limiting factor as they believe in and stick to supply management instead of demand management as a way of addressing water distribution issues. Problems of groundwater pollution and overexploitation causing saltwater intrusion are not included in the problem formulation by the public. In terms of *strategies and instruments* we observe that fragmentation exists between urban planning and water resource management. Water quality aspects are insufficiently included in water resource management. There is a lack of integrated planning and integrated water resources management (IWRM). A strategy for demand management is missing, since the prevailing strategy is mostly supply oriented. In terms of *distribution of resources* we observe that strong property rights on agricultural water resource use exist. Tourism and urban development often require more water than is available. The Marina Baja Water Consortium now forms an important part of the governance regime related to the trading of used water, and improving water use efficiency (reallocation among users). Water users are not yet fully bearing the costs of infrastructure, exploitation and externalities. General budgetary funds of various administrations are still involved.

Contextual (institutional-based) factors

The case of the Marina Baja demonstrates an interesting adaptive response in terms of constructing a new actor arrangement at the local level between municipalities and private stakeholders (farming community and the tourism branch) for the purpose of redistributing resources among these actors. The drought in 1978 created a strong *sense of urgency* and a credible threat of future water deficiencies that could occur if nothing were done (*mutual dependencies*). The prospect of growing demand functions as a joint problem, while the possibility of water reallocation functions as a joint opportunity (*shared cognitions*). In terms of *shared values* it is remarkable that the stakeholders still rely on water reuse, inter-basin transfers and desalinization as their approaches to managing water supply. They believe these options are able to satiate the increasing water demands from agriculture, domestic use and tourism without the need for demand restrictions. In that way, property rights do preserve existing modes of water management and water utilization but compensation payments are introduced to better balance the rival uses of a limited stock of water resources.

13.4 France (Hérault river basin)

The Hérault is a mountain-fed river basin in the South of France. It reaches the Mediterranean Sea near Agde, west of the city of Montpellier, and has a length of 148 km. The Hérault basin is situated in the region of Languedoc-Roussillon.

Water utilization pattern

The Hérault river basin suffers from sanitation problems, as during times of low flow the amount of effluent from wastewater treatment plants has a very high impact on the water quality of the river (compared to periods of high river flow). This is partly the result of *natural conditions*. Water shortage periods in summer constrain irrigation for a few weeks, but rarely result in a ban on crop irrigation. Supplies have not been low enough to affect drinking water supply, industrial use and hydropower in summer. Similarly, there is no pattern of increasing drought. On the contrary, very violent flash-floods can happen in the autumn and winter. However the awareness of issues related to water deficiency are increasing due to *socio-economic conditions*. There is a changing population due to new settlers from outside the area who now represent the majority of the inhabitants. The demand itself is not changing, but the sensitivity of new citizens to irrigation affecting recreational use options is. Tourism is an increasingly important source of income for the area and is thus playing a more important role in discussions on ways for determining water distribution.

Policy responses

As a result of their perceived water stress, government agencies in the Hérault area have promoted the principle of collaborative management and planning of water resources. They have used a planning instrument – the SAGE procedure (Schémas d'Aménagement et de Gestion des Eaux; in English: Integrated Water Management Plan at River Basin Level) – which was defined at the national level by the 1992 French Water Act. SAGE creates a platform where stakeholders can debate and negotiate. Applying this model means a change from the traditional top-down model to a more flexible participatory model. This can be identified as an *adaptive response*. SAGE is one of the major innovations of the 1992 Water Act. The SAGE process aims at including all water issues and all relevant stakeholders in a river basin. The process results in active participation and broader attention is given to river-related concerns as compared to solely urban and economic growth concerns.

Governance regime

In terms of *levels and scales* we observe government agencies that traditionally work at the level of the department, now have to cooperate at the basin level. This strengthens the cooperation between the central and the department level in terms of co-governance. Although public agencies are supposed to join forces, not all of them support the SAGE planning procedure. In terms of *actors and networks* we observe that SAGE is a drastic change in the sense that stakeholders such as environmental associations, canoe rental companies, etc. can contribute to policymaking. However, not all of the stakeholders are well represented. Legitimacy issues have arisen since government agencies select the representatives and since only organized stakeholders are represented in the process. Consultation is very formal, and the rather technical language used by the expert community results in the general public not being involved. In terms of *perspectives and objectives* we observe that it is difficult to achieve joint problem formulation. Traditional stakeholders refuse to accept the existence of water stress due to the inclusion of new values. As a result, public agencies avoid speaking about water conflicts and consider them only as possible future tensions. In terms of *strategies and instruments* we observe that SAGE applies to the new paradigm of IWRM. In terms of *distribution of resources* we observe that representation of stakeholders is missing 'legitimacy', which is an important resource for making policies effective. The SAGE procedure thus does not provide stakeholders with sufficient trust, even though the purpose of the procedure is to gain more trust and support from the public.

Contextual (institutional-based) factors

In the Hérault case the SAGE procedure plays a crucial role in providing an adaptive response to water deficiency. The procedure restructures the cooperation between administrative levels, leads to the involvement of new actors, reformulates the policy problem and results in more integrated management. The *sense of urgency* comes from anticipation of future threats of water shortage and from present rivalries. There is no urgency related to immediate water scarcity. In terms of *shared values*, a tradition of cooperation developed through SAGE brings stakeholders together. In terms *of mutual dependencies* among stakeholders, it is interesting to have a closer look at the role of property rights in the Hérault case. Rivalry exists between farmers (irrigation) and stakeholders making indirect use of water (fishing, swimming and canoeing). Fishing associations argue that farmers do not comply with the 1994 Fishing Act, which requires minimum flow. They also argue that farmers could prevent water stress by changing irrigation practices (for example by switching to drip irrigation). Farmers argue that they have historical legitimacy for water use in the basin and that they should not have to accede this position to new environmental concerns. Political pressure is often used to oppose cases that are taken to court. The social pressure on farmers has developed to a credible alternative threat and has led them to adopt a more adaptive response of exploring the possibility of constructing small reservoirs. Financial institutions, such as the Rhône-Méditerranée-Corse Water Agency, support conflict resolution by subsidizing studies and investments to improve irrigation efficiency. In this section, political pressure, social pressure and financial institutions are shown to function as 'institutional interfaces' and 'brokers' that help to achieve negotiated compromises and to avoid continuous litigation.

13.5 United Kingdom (Nene river basin)

The river Nene is located in the East of England. It rises from three sources in the county of Northamptonshire and ends at the North Sea coast. It is the tenth longest river in the UK (142 km). It is a tidal river whose flow and level are influenced by the tides of the North Sea, and as such is affected by problems of saltwater intrusion. The inland flow of saltwater from the sea limits the consumptive uses of the river water.

Water utilization pattern

In terms of *natural conditions*, the Nene river basin is an area with some of the lowest mean rainfall across the country. Although it is not a semi-arid area, the catchment does experience episodic drought events, exacerbated by low rainfall levels in the catchment. Meanwhile, the catchment also experiences significant floods, the most recent of which occurred in the spring of 1998. In terms of *socio-economic factors* abstractions from the river constitute 93 per cent of water supply and is as such, considered to be 'fully abstracted'. Low summer flow coupled with high public water services demand can cause abstraction problems. Additionally, due to a rising population (and a government policy to develop 400,000 new homes in the period 2005–2025), more new housing is inevitable. These predictions coupled with low rainfall means that the area could become water stressed (not permanently, but episodically).

Policy responses

The 1975/1976 dry years resulted in the Ministry of Environment declaring a national drought and that public standpipes would be the only source of public supply. This naturally induced event led to changes that were less structural and more procedural, such as the review of drought warning systems in place and more public information campaigns. The warning procedure requires that abstraction automatically cease when demand flow drops to a level that threatens the environmental health of the river. In

1995, the water authorities, industry and agriculture experienced low-flow impacts. Domestic use was, however, less impacted than in the previous drought in 1975/1976. Since then, industry and agriculture have adopted more efficient water use procedures. Once the drought events were over, various bodies looked to improve their own management processes and technical capacity, rather than view these events as a social or political/economical issue. After water levels and water services returned to normal capacity, the agenda moved to 'what we could do better next time' (in terms of crisis management). No policy changes developed around preventing a drought or structurally seeking to change patterns of water use. The relative cost of water did not rise as a consequence of these droughts. Investment, research and development efforts were focused on technical- and engineering-based solutions such as better forecasting, enlarging reservoir capacity, reducing water leakage within the infrastructure and developing new reservoirs. The local experts that were interviewed believed that reducing leakage rates and encouraging the public to take up water efficient behaviours would be counteracted by the pressure of the rising population and associated proposed housing, and hence placing greater pressure on the freshwater management regime in the catchment. To conclude, we found no real adaptive responses but only *reactive responses* to drought in the Nene case.

Governance regime

In terms of *levels and scales* we observe that the 1989 privatization of the water sector provided for a market-orientated regime, which seeks no funding from central government and is run along full cost recovery lines. At the regional level, a private company is responsible for water management as a commodity, not as a resource. At the national level, the Environmental Ministry governs resource management. The 1995 drought took place in the post-privatization era. Each of the water companies involved were negotiating tactics individually and lobbying the Environmental Ministry to balance their statutory public water supply duties with the licences held by agriculture and industry. The public sector (the national government) considers droughts as the management responsibility of private companies, while the private sector operates under the premise that it can negotiate to broaden their licences during drought periods. Due to the fractured nature of the governance regime, the concept of water scarcity is not currently on the national government's agenda. In terms of *actors and networks* we observe that there is no participatory process at the regional scale where the public can participate. They can only participate through the national voluntary CAM-process (Catchment Abstraction Management Plan). This process can be viewed as a forecasting method to assess potential conflicts amongst water users (as stakeholders) in the near future. The CAM-process is projected to be reviewed on a six-year cycle and to provide conclusions and recommendations regarding the process. There is no mechanism for price regulation by regional stakeholders (as consumers). In terms of *perspectives and objectives* we observe that the public image of the country is one of sufficient water resources with a dominant cultural belief that there is no water scarcity in the UK. The general perception amongst the public that freshwater is of good quality and not in short supply, may be an explanatory factor for the slow uptake in recycling technologies and water reuse systems within industry and agriculture. Water issues are not a national debate as there is no resource perception and people trust in resource availability and management (engineering of water services; warning systems) to provide for their water needs. Water use reduction is not considered as an option for consumers. Floods and droughts are generally considered to be episodic rather than structural and are thus not seen as the result of socio-economic factors, but as the result of poor management decisions. Stakeholders mainly trust engineering and forecasting to provide solutions and as such only a radical shift in the amount of available water is likely to reduce water consumption. In terms of *strategies and instruments* we observe that although there is integration of water management and environmental management at the national level, a strategy for water resource protection at the regional level is missing. In terms of *distribution of resources* we observe that the post-privatization mode of regulation has prevented price swings (due to the role of the national price regulator). Water companies in the UK are not satisfied with the available flexibility of cost recovery mechanisms however, and they argue that they do not have enough funds to reinvest in improving infrastructure. This could encourage the government to reassess price capping policies that had been intended to set limits on the prices charged for water and sewerage services, while taking into account proposed capital investment schemes and expected operational efficiency gains. Resource protection is not included yet in the price of water services. Setting quotas only makes sense when water supply companies can manage supply more efficiently. Demand management and resource management are not in their hands.

Contextual (institutional-based) factors

Severe droughts in 1975/1976 and 1995 created a limited and short-term sense of urgency in the UK and the Nene area. However, policymakers responded in a reactive manner. We did not find any policy measures or policy plans that anticipate an increasing demand following the future construction of 400,000 new homes in an arid area. The dominant shared values related to unlimited water use, the *dominant shared cognitive perception* that water management is an engineering issue, and the strong role of the private sector (in terms of *mutual dependencies*) explain the preservation of the existing mode of water management.

13.6 Netherlands (Maas river basin)

The Maas (or Meuse) is a major European river, rising in France and flowing through Belgium and the Netherlands before draining into the North Sea with a total length of 925 km. The Maas Works is a river restoration plan that aims to restore floodplains and natural river dynamics along a 220 km stretch located in the south of the Netherlands contained by the provinces of Limburg and North-Brabant.

Water utilization pattern

In terms of *natural conditions*, the Maas valley is not an arid or semi-arid area, but suffers from water depletion due to the fast water passage of a large river in a narrow river basin. The Maas valley suffers from water depletion and flooding, which are affected by a loss of space in the river basin, particularly within its floodplains. The combination of a narrow river basin and rival land uses in the floodplains has led to insufficient river valley space for natural river dynamics. In dry periods, the catchment suffers from water depletion. In periods of heavy rainfall, the river basin suffers from flooding due to excessive rainwater coming from France and Belgium. Heavy river floods in 1993 and 1995 resulted in a high sense of urgency due to extensive material damage and the evacuation of 8,000 people. In terms of *socio-economic factors*, gravel extraction (on a concession basis) in the floodplains has degraded the river landscape and caused sound and vibration nuisance for the inhabitants of the Maas valley. Water drainage systems on agricultural land along the river and the resulting decline of the groundwater table have caused water depletion and degraded the quality of the ecosystem in floodplains. Urban and industrial estate expansions in the river floodplains have destroyed the ecosystem and reduced the chances for river dynamics and water storage. Conditions for shipping transport on the Maas have worsened due to a lack of investment in navigation-infrastructure. Drinking water production from groundwater sources has lowered groundwater tables, thereby affecting water depletion.

Policy responses

The Maas is selected as a case study because of the *adaptive response* developed to handle both water depletion and flooding problems by means of an institutional arrangement in which economic exploitation of floodplains is used for the cost recovery of floodplain restoration. During the period 1985–1992, the Maas Works restoration plan was developed to restore natural dynamics where possible, improve ecological conditions and also to improve shipping conditions with new investment in a navigation route. Meanwhile, plans were initiated to move agricultural operations beyond the floodplains to allow a rise of groundwater tables, to stimulate the water supply sector to gradually shift the production from groundwater to surface water and to set limitations on gravel extraction by allowing more public participation in the attribution of new concessions. Severe river floods in 1993 and 1995 created the political will to develop more storage capacity in the river valley by widening and deepening the riverbed. As a result, new elements were added to the national governments' restoration plan (the Maas Works). These elements provide for an expansion of gravel extraction as a means to recover the costs of restoration works. Municipalities were also expected to use their land-use decision-making authority to prevent further urban and industrial expansions in floodplains and agricultural operations would be further encouraged to move beyond the floodplains. The population was also to become more involved in the process of creating space along the river. Transboundary cooperation and fine-tuning across the whole river basin would be undertaken to better control and anticipate the water flow through the Maas.

Governance regime

In terms of *levels and scales* we observe that although the Maas is an international transboundary river, the Netherlands considers it to be a national river basin and not a regional river basin. This makes it the responsibility of the national government as opposed to the watershed based Water Boards which are separate government bodies generally responsible for regional water management issues. The national government does not want to invest national money into the Maas restoration plan as they believe that the provincial government should recover the costs of the Maas restoration with revenues from the gravel extraction concessions. The provincial government argues that the 1993 and 1995 floods were national disasters. Citizens in the province of Limburg who are opposed to gravel extraction consider the expansion of gravel concessions to finance the Maas restoration as an unfavourable solution. On the other hand, the national government maintains that it is a sustainable solution at the provincial scale for the water basin. In terms of *actors and networks* we observe tensions between the gravel extraction concession holders on the one side and the local population on the other side. The gravel extraction group is supported by the national government in its aim to expand gravel extraction, whereas the local population is supported by the provincial government in its aim to reduce gravel extraction and get the Maas restoration subsidized from the national budget. Furthermore, the Maas restoration plan created a rivalry between farmers who are afraid of losing agricultural land for the purpose of floodplain restoration, and the national and provincial authorities who are united in their aim to create 'room for the rivers'. In terms of *perspectives and objectives* we observe that the national government's requirement of budget neutrality (full cost recovery for the Maas restoration plan out of revenues from gravel extraction), brought an alternative into the discussion on the Maas restoration plan. The citizens suggested an option of not restoring the Maas and setting up an arrangement for damage compensation in case the Maas area is confronted with new flooding. In terms of cost effectiveness this could be a better option for the local population, although this option is less legitimate and credible. The flood risk estimations and necessary prevention levels associated with the Maas restoration plan continuously changed due to uncertainty about the future impact of climate change, in combination with the requirement of budget neutrality. The associated scenarios were changed based on the amount of gravel extraction that would be needed to recover the costs associated with the flood prevention measures. This continuous uncertainty caused much delay during the development and implementation of the Maas restoration plan. In terms of *strategies and instruments* we observe that the adaptive responses included within the Maas restoration plan make use of the possible link between flood defence, river restoration and gravel extraction as a window of opportunity. This arrangement is a potentially effective arrangement for adapting prevailing resource governance structures to an intensive water utilization pattern, such as was observed in the Marina Baja Water Consortium in the Spanish case. In terms of *distribution of resources* we observe that the success of adaptive responses in the Maas river basin very much depends on the cooperation of land owners in floodplains (like farmers) and the general public, as well as on cooperation with municipalities and industrial stakeholders (like gravel extractors). The property rights of gravel extractors and the requirement of the national government for budget neutrality result in a transfer of revenues from one natural resource (gravel) to the restoration of the river as another natural resource.

Contextual (institutional-based) factors

The 1993/1995 floods increased the *sense of urgency* for the Maas restoration plan. This plan is seen as an adaptive response to a complex set of problems, including water depletion in the river basin. In terms of *shared cognitions*, the Maas restoration plan functions as a joint opportunity for stakeholders to make effective use of the possible link between flood defence, water depletion, river restoration and gravel extraction. In terms of *mutual dependencies* between actors, the national requirement for a balanced budget forces regional stakeholders to cooperate on a compensation arrangement between public and private actors. Not restoring the Maas and setting up an arrangement for damage compensation in case the Maas area is confronted with new flooding is a credible alternative threat, since the compensation costs should be covered by the provincial authorities. It is the lack of *shared values* (disputes on the consequences of the Maas restoration for the position of agriculture and gravel extraction in the floodplains, and disputes over the national ownership of the regional Maas problems), which has delayed the implementation of the Maas restoration plan.

13.7 Discussion and conclusions

In four case studies across South and North Europe we investigated how adaptive policy responses evolve from the underlying institutional setting (governance regime) and are triggered by contextual factors. Adaptive policy responses can develop based on a deliberate concept transfer from one region to another or as a learned response to a series of threats accompanied by the recognition of the expected continued experience of future threats. The results of this research show examples of where adaptive policy responses developed (or not) under different circumstances. In the Hérault case and the Maas case we found examples of adaptive responses in the form of *integrated planning*. Integrated planning is based on the scale of the river basin and crosses administrative boundaries. It takes into account rivalries and functions as a tool to organize participation and commitment of all relevant stakeholders. It preserves the status of water values that are otherwise rather weakly protected and it provides a coordinated policy approach for the various instruments that tackle different parts of the policy problem. In the Marina Baja and Maas cases we found examples of adaptive responses in the form of *compensation arrangements*. Such arrangements help to overcome rivalries in case of resource scarcity, which includes scarcity of space for water storage. Through this solution, some actors helped others by transferring institutional resources such as money. In the Nene case especially, we found evidence that reliance on sufficient water supply by means of engineering solutions causes actors to resist demand restrictions. Such an attitude, results in the preservation of existing modes of water management and thus in more reactive responses.

Contextual factors

As is discussed in Chapter 3, contextual factors exert an influence on policy and implementation processes (and their effectiveness) through their influence on the characteristics of the actors involved. We distinguished between natural factors, socio-economic factors and institutional factors. *Natural factors* include physical characteristics of the river basin or catchment area as well as weather- or climate-related events such as droughts and floods. Droughts (Marina Baja 1978) and floods (Nene 1998; Maas 1992 and 1995) can create a high sense of urgency, which is influenced by the way in which these events are perceived by the public. Relevant *social-economic factors* are considered to include utilization patterns and water demands. The cases show that the increasing demand for agriculture (Marina Baja, Maas), domestic use (Nene, Marina Baja, Maas) or tourism and recreation (Marina Baja, Hérault) result in rivalries and a debate on water as an unlimited commodity versus water as a limited resource. Water transfer from one aquifer to another (Marina Baja, Hérault) raises questions about where to draw river basin or catchment boundaries and how to define the scale at which the stock of the resource is limited. Property rights on urban and agricultural land use (Nene, Maas) have led to restrictions of river dynamics and thus affect the storage capacity and natural function of the river basin. Existing modes of water management and water utilization can be preserved due to intact property rights and thus maintain rivalries. On the contrary a lack of property rights is experienced in the case of recreational use (Hérault, Nene) and ecology (minimum flow and river dynamics) as new values (Marina Baja, Hérault, Nene and Maas).

Does gradual degradation of the resource (instead of sudden events like droughts and floods) trigger policymakers to act in anticipation of a future threat? The Marina Baja case demonstrates the presence of anticipation by local policymakers about growing demands for agricultural, domestic and tourist uses as they develop solutions based on water reuse, transfers and desalinization (though no demand restrictions were proposed). The Hérault case demonstrates anticipation of a threat from water transfer from the aquifer to another region (a dispute which was not solved during our period of research). The Nene case does not show policymakers anticipating domestic demand issues related to the future construction of 400,000 new homes. The Maas case shows that anticipation of new flood threats slowed down due to implementation problems with the flood defence policy. Overall, we were able to distinguish the presence of *institutional factors* like shared values, shared cognitions and mutual dependencies affecting the behaviour of the stakeholders involved and discovered how these aspects improved the chances for a more adaptive response in the different cases.

Governance regime

Policy responses develop partially based on the underlying institutional setting or governance regime. We used five dimensions of governance to highlight the various avenues through which the concept of adaptive policy response arises under different governance regimes. In terms of *levels and scales* the interconnection and coherence of central/national policymaking (with top-down guidance) and decentralized arrangements was seen to be influential. Decentralized arrangements could be bottom-up arrangements of stakeholders at the local or regional level (like drainage boards), or could be resource authorities at river basin scale. In order to support adaptive policy responses to water scarcity, stakeholder organizations should include those that perceive water as more than a commodity, and hence include other use functions of the resource. In general, multilevel governance should prevent levels from functioning in an isolated manner and should interconnect top-down and bottom-up approaches. In terms of *actors and networks*, adaptive responses were influenced by the presence of rival values and whether users have an opportunity to be represented or involved in decision making on river basins and water catchments. Planning functions can provide an institutional interface that creates opportunities for representation since water policy communities are often closed expert communities sharing highly professionalized language. The problem formulation determines which actors are considered as stakeholders. Involving new stakeholders thus depends on reformulating the problem as well as the receptivity of a policy community of experts to have an open mind towards new water values and to create access for outsiders with an interest (Jeffrey and Seaton, 2004; de Boer and Bressers, 2011). Involvement of stakeholders at the local and regional scale requires bottom-up processes. In terms of *perspectives and objectives* discourses that exist around present and future water problems were observed to influence the process of adaptive policy responses. Discourses that support sole reliance on engineering solutions for addressing water scarcity and limit attention to the importance of related social dilemmas, and water management that only considers physical problems that can be solved with technical interventions were not observed to support adaptive policy responses. Integrating new values into the existing water culture in a country or an area is something that deserves attention in developing an adaptive policy response. In terms of *strategies and instruments* drought and water deficiency problems were affected by the development and discussion of a strategy for demand management. Water management and land-use management were observed to be co-dynamic processes and when integrated can address issues related to scarcity. Property rights can preserve existing modes of water utilization and thus approval procedures for zoning and land-use plans that are screened for water effects by water authorities can be helpful in developing more adaptive perspectives by policymakers. In terms of *distribution of resources* adaptive policy responses are favoured when users pay for the full costs of water services and when externalities are internalized into the price of water. Fixation of a water demand quota and creation of tradable quotas are examples of adaptive policy responses that were observed to stimulate efficiency on the side of water users.

This overview summarizes the important contextual factors that were observed to be related to the development of adaptive policy responses to water scarcity issues in the four case studies researched. A conversion towards more adaptive policy initiatives begins with changes on each of the five dimensions of the underlying governance regime (Bressers and Kuks, 2008). Although contextual factors (natural conditions, socio-economic factors and institutional-based factors) play an important role in determining if institutional change will happen, some suggestions are provided through the discussion related to how institutions can more actively improve the ability of a governance regime to adapt.

References

Aquadapt (2003) European Commission, 5th Framework Programme, RTD project under the thematic programme 'Energy, Environment and Sustainable

Development' Key Action 1 'Sustainable Management and Quality of Water' www.ua.es/es/internacional/internacionalizacion/aquadapt, accessed 15 October 2012.

Bressers, H. and Kuks, S. (eds) (2004) *Integrated Governance and Water Basin Management: Conditions for regime change and sustainability*, Kluwer Academic Publishers, Dordrecht/Boston, MA/London.

Bressers, H. and Kuks, S. (2005) 'Integrated water regimes and sustainable use: Learning from comparing cases' in M.A. Lange, D. Poszig, and A. Herrmann (eds) *Water on Mediterranean Islands*, ZUFO, Munster.

Bressers, H. and Kuks, S. (2006) 'Water basin regimes in Europe and institutional conditions for their sustainability', in V.I. Grover (ed.) *Water: Global common and global problems*, Science Publishers, Enfield, NH.

Bressers, H. and Kuks, S. (2008) 'Institutional sustainability – Integrated regimes for river basin management: What helps them grow and do they work?' in A. Dehnhardt and U. Perschow (eds) *Sustainability in River Basins: A Question of Governance*, Oekom Verlag, Munich.

Bressers, H., Fuchs, D. and Kuks, S. (2004) 'Institutional resource regimes and sustainability' in H. Bressers and S. Kuks (eds) *Integrated Governance and Water Basin Management*, Kluwer Academic Publishers, Dordrecht/Boston, MA/London.

Directive 2000/60/EC of the European Parliament and of the Council of 23 October 2000 establishing a framework for Community action in the field of water policy.

de Boer, C. and Bressers, H. (2011) *Complex and Dynamic Implementation Processes: Analysing the Renaturalization of the Dutch Regge River*, University of Twente and Dutch Water Governance Centre, Enschede and the Hague.

de Boer, C., Bressers, H. and Kuks, S. (2011) 'Coordination of policies and governance regime requirements in Dutch freshwater management', *Policy Quarterly*, vol. 7, no. 4, pp. 3–9.

Jeffrey, P. and Seaton, R.A.F. (2004) 'A conceptual model of "receptivity" applied to the design and deployment of water policy mechanisms', *Environmental Sciences*, vol. 1, no. 3, pp. 277–300.

Kuks, S. (2005a) 'Governance Regimes for Responsive Water Management: Part 1 – Comparative analysis of four European regimes (Spain, France, UK, Netherlands)', research paper presented at the ARID Cluster Conference 'Drought and water deficiency: from research to policy making', 12–13 May 2005, Cyprus.

Kuks, S. (2005b) 'Governance Regimes for Responsive Water Management: Part 2 – Policy implications to enhance the adaptive potential of water governance regimes', research paper presented at the ARID Cluster Conference 'Drought and water deficiency: from research to policy making', 12–13 May 2005, Cyprus.

Pahl-Wostl, C., Sendzimir, J., Jeffrey, P., Aerts, J., Bergkamp, G. and Cross, K. (2007) 'Managing change toward adaptive water management through social learning', *Ecology and Society*, vol. 12, no. 2, art. 30.

White, C. (2012) 'Understanding water scarcity: Definitions and measurements, GWF Discussion Paper 1217', Global Water Forum, Canberra, Australia.

14 The transfer of Building with Nature approach in the context of EU Natura 2000

Vera Vikolainen, Kris Lulofs and Hans Bressers

14.1 Introduction

This chapter contains two types of transfer: the transfer of the European Union (EU) Natura 2000 biodiversity policy from the supranational level to the member-state level and the transfer of Building with Nature, an emerging approach to water infrastructure design and development. The two types of transfers are related: the transfer of the EU Natura 2000 to a large extent determines the context for the transfer of the Building with Nature approach within the EU.

The EU Natura 2000 biodiversity network is a policy strategy for the implementation of the EU Birds and Habitats Directives (EEC, 1979, 1992). The directives introduced a protection regime for species and habitats while the designation of Natura 2000 areas had substantial consequences for plans and projects with potentially negative effects on the ecological values of species and habitats. As the transfer of Natura 2000 policy involved the imposition of a certain policy onto member states' governments by an external government or supranational institution it is an example of a *coercive transfer* (Dolowitz and Marsh, 1996). The implementation of Natura 2000 and the directives' requirements with respect to water infrastructure was problematic and caused project delays and cancellations across north-west Europe (Van Hooydonk, 2006; Mink, 2007).

Building with Nature is a modern water management concept also referred to as Eco-dynamic Design and Development, or Working with Nature (PIANC, 2011; EcoShape, 2012). Building with Nature aims to integrate ecosystem characteristics and socio-economic objectives at the initial design stage of water infrastructure projects in estuaries and coastal zones. This approach proposes to use nature's status and dynamics as a starting point to create designs where new infrastructure is developed alongside improved ecosystems. The initial expectation of Building with Nature proponents was that the EU network of protected Natura 2000 areas would have a negative effect on the transferability of this approach within the EU. The reason for this is that Natura 2000 aims to protect existing characteristics of ecosystems, while Building with Nature aims at ecosystem development, which makes physical interference necessary to realize the desired potential of ecosystems. Building with Nature experiments are ongoing in the Netherlands (e.g. large-scale sand nourishment of the Dutch coast as means of flood protection, also known as the 'Sand engine') however there are not yet any completed examples of the transfer of this water management concept to other EU member states. Nevertheless, the approach is already being promoted by the European Commission in the area of port development and dredging (European Commission, 2011) and inland waterway development and management (European Commission, 2012). Hence it is helpful to anticipate the chances of Building with Nature succeeding within the varied context of Natura 2000 in different EU member states. This attempt to assess the effect of an initiative before it is put into place is also referred to as *prospective transfer evaluation* (Mossberger and Wolman, 2003). Prospective transfer evaluation of Building with Nature is particularly interesting given the fact that this approach has a potential to become a 'best practice' for water management in estuaries and coastal zones.

This chapter is structured around two research questions. The first research question is descriptive: what does the transfer of Natura 2000 in north-west EU member states look like? To answer this question, the chapter presents the review of literature on Natura 2000 implementation in north-west Europe, including France, Belgium, the Netherlands, Germany, and the United Kingdom (UK). These member states face similar challenges in implementing the directives due to their high population density and intense economic activity. However, the other EU member states (Austria, Bulgaria, Cyprus, Czech Republic, Denmark, Estonia, Finland, Greece, Hungary, Ireland, Italy, Latvia, Lithuania, Luxembourg, Malta, Poland, Portugal, Romania, Slovakia, Slovenia, Spain, Sweden) are also included when they are covered in the literature referred to in this chapter. Extra attention is paid to the experience in the Netherlands because it is the member state where Building with Nature experiments currently take place.

The second research question is exploratory: what are the opportunities for successful transfer of the Building with Nature approach within the EU Natura 2000 context? To answer this question the chapter builds on the understanding of structural and specific contextual factors as specified in the Contextual Interaction Theory (CIT) (Chapter 3 of this book). CIT is used as a conceptual lens to structure the Natura 2000 context into a manageable framework, but not in the strict sense of the word 'theory' for hypotheses generation and/or testing. With the help of the CIT, the specific and structural context factors, which constitute the Natura 2000 context, are identified and systematized. Understanding of the Natura 2000 context is a first step in the prospective evaluation of Building with Nature transfer and could guide the actual transfer of this concept across the EU in the future.

The chapter is structured as follows. In Section 14.2 the historical background of Natura 2000 transfer is sketched based on the national implementation arrangements made by the member states to fulfil the requirements of the Birds and Habitats Directives. In Section 14.3 Building with Nature as a new approach for the design and development of coastal and delta infrastructure is introduced. In Section 14.4 the framework for prospective transfer evaluation is offered and the potential for Building with Nature to become a best practice is discussed. In Section 14.5 the two research questions are answered.

14.2 Implementation of the Birds and Habitats Directives

Nature conservation policy in the EU is based on the Birds Directive (79/409/EEC) and the Habitats Directive (92/43/EEC). A directive is one of the formal legal instruments for developing EU policy. The provisions of a directive are binding regarding the results to be achieved. For example, 'favourable conservation status of species and habitats' under the Birds and Habitats Directives is binding upon each member state to which these directives are addressed. The aim of a directive is to bring together and coordinate the laws of the member states in the policy field addressed by the directive. For each directive, a deadline is set to harmonize national laws in a way that secures the achievement of a directive's goal. Below we briefly introduce the main goals as set by both the Birds and Habitats Directives, as they establish and shape the governance of Natura 2000 areas.

The Birds Directive aims to conserve all species of naturally occurring birds in the wild state in the European territory of the member states, whereas the Habitats Directive aims to contribute towards ensuring biodiversity through the conservation of natural habitats of wild flora and fauna in the European territory of the member states (Article 1 of the Directives 79/409/EEC and 92/43/EEC, respectively). The member states were required to take the following steps in order to achieve the goals set by the directives:

1 Transpose the directives into national law (within two years of its notification).
2 Select and designate protected sites: Special Protection Areas (SPA) prescribed by the Birds Directive and Special Areas of Conservation (SAC) prescribed by the Habitats Directive.
3 Take conservation measures in the designated areas involving, if need be, a management plan.
4 Assess any plans and projects that are likely to have significant effect on the designated areas according to the procedure of Article 6 of the Habitats Directive.

In 2010, all member states had a legal framework for the implementation of the directives and the designation of SPAs and SACs into Natura 2000 sites, while more than 80 per cent of the member states have selected and designated about 80–100 per cent of their sites. Most of the member states plan to complete formal SAC designations in 2013–2014 (Van Apeldoorn *et al.*, 2010). The formal implementation stage is underway and its progress has been analysed in the literature.

Transposition of the Birds and Habitats Directives into national legislation

The way in which national legislation transposes the directives partly depends on the constitutional structure of the country. In France and the Netherlands, the central authorities passed the necessary legislation; in Belgium, it was passed at the regional level; in Great Britain legislation applies to England and Wales, while Scotland and Northern Ireland have their own legislation. In the Netherlands, species and habitats protection were transposed with two separate laws: site protection in the Amended Nature Conservation Act 1998 (entered into force in 2005) and species protection in the Flora and Fauna Act (2002). In the Belgian region of Flanders, habitat protection was transposed in the Amended Nature Conservation Decree 2002 (entered into force in 2004) and species protection is still in development. In the UK, the Birds Directive is transposed in the Wildlife and Countryside Act of 1981, the Wildlife Order (Northern Ireland) of 1985 and the Nature Conservation and Amenity Lands Order (Northern Ireland) of 1985. The Habitats Directive, including species protection, is transposed in the UK Habitats Regulations of 1994. Austria, Germany, England, Flanders and the Netherlands transposed the directives' provisions in nature conservation legislation. In France and Sweden, the transposition has taken place in environmental law: Environmental Code (statutory provisions) and Rural Code (regulatory provisions) in France and Environmental Code in Sweden. The legislation in countries with a federal structure, like Germany and Austria, varies widely on details in the individual federal states.

In many member states, the directives were not transposed within two years of their notification (the deadline set by the Commission). This could have been caused by the member states not prioritizing and allocating adequate resources to the implementation of the directives. At the same time, many environmental non-governmental organizations (NGOs) could offer resources and expertise required by the directives and pressured member state governments and administrations by making complaints of non-compliance to the Commission (Paavola, 2004). Such complaints are one of the ways for the Commission to stay informed about the process and to fulfil its function as a watchdog over the implementation of the EU treaties. In accordance with Article 226 of the EC Treaty, the Commission can enforce sanctions against member states for non-compliance with the directives by referring the cases to the European Court of Justice. The European Court of Justice issued more than sixty rulings concerning the Birds Directive and forty-five rulings concerning the Habitats Directive (Beijen, 2010). Most of them were infringement procedures initiated by the European Commission for incorrect, incomplete or late transposition into national legislation, for instance in the Netherlands, Belgium, France and Italy.

Where the Habitats Directive was not correctly or timely transposed, the cases pending before national courts were checked directly against the directive's provisions (Backes *et al.*, 2006). Furthermore, the areas that were included on the list of Sites of Community Importance (SCI) but not yet designated nationally (this could take up to six years), were already protected under Article 4 paragraph 5 of the Habitats Directive. Article 6 of the Habitats Directive applied to these areas directly, irrespective of whether this provision has been correctly transposed into national law. Direct application of the Habitats Directive compelled by national courts led to many problems in practice. The situation in the Netherlands is a good example in this regard.

In the Netherlands, the Birds and Habitats Directives were completely transposed into national legislation only in 2005. The Dutch government failed to inform actors about the consequences of both directives. This failure caused a legally obscure situation and led to many problems (Beunen, 2006). The actors involved in the decision-making processes, such as project developers and street-level officers simply did not know about these directives or did not know how to apply them. This lack of attention, knowledge and awareness led to a decision that rarely met the requirements of the directives, i.e. an appropriate assessment of project effects was not done (Beunen, 2006; Beunen *et al.*, 2009). As a result, many plans and projects were challenged in court by environmental NGOs because of the presence of protected species, and thus protected species became a symbol of the directives' restrictive effect on spatial development in the Netherlands (van der Zouwen and van Tatenhove, 2002). After the first cases were decided upon in court, the role and influence of courts, judges and lawyers increased, as did the dominance of legal perspective on the directives' implementation in the Netherlands. Many of the articles published about the consequences for Dutch planning practices have been written by lawyers (e.g. Bastmeijer and Verschuuren, 2003, 2004; Verschuuren, 2003, 2004). Initial negative experiences with the directives have created an image that economic developments in the Netherlands are 'locked up' because of the implementation of these two directives. To some extent, this image is still influential in current political and professional debates (Arnouts and Kistenkas, 2011).

Natura 2000 site selection and designation

The Birds Directive gives the member states limited freedom in selecting the protection areas. It only allows the use of ornithological criteria to designate SPAs and only the reasons of public safety (e.g. flood risk) to derogate from these criteria. Site protection under the Birds Directive (Article 4 paragraph 4) is extremely strict and does not allow any exceptions. This regime was later replaced by the habitat assessment (Article 6 paragraphs 2–4 of the Habitats Directive), which is more flexible, as it allows derogations for reasons of overriding public interest. Species protection is extremely strict and applies to all birds occurring in the wild (not only to endangered species) and derogations are limited. Similarly, only ecological criteria could be used to designate SACs under the Habitats Directive. Strict designation requirements of both directives explain the unwillingness of the member states to designate SPAs and SACs (Beijen, 2010).

Furthermore, scientific information for site designation was not always available and the directives provided no detailed designation guidelines. As a result, several member states have designated too few areas, areas that are too small, or have later tried to change their initial designations. This led to a considerable number of rulings of the European Court of Justice on the failure of member states to submit lists of designated sites in conformance with the deadlines (Paavola, 2004). Recent statistics show that the member states designated between 3 per cent and 26 per cent of their total national area as SPA and between 7 per cent and 31 per cent of their total national area as Sites of Community Importance (SCI, soon to be designated as SAC) (Natura 2000 Barometer, 2010).

The comparison of designated sites in Flanders, the Netherlands, England and the German federal states Lower Saxony and North-Rhine Westphalia shows that the Netherlands has designated relatively large areas: half of the sites are 100 ha or larger (Bouwma *et al.*, 2008). Flanders, the Netherlands, Lower Saxony and North-Rhine Westphalia designated an almost equal percentage of their total surface area (between 8 per cent and 14 per cent). England had the least amount at 6 per cent and the UK total is 8.9 per cent. The Netherlands has relatively many SPAs under the Birds Directive due to the importance of its territory for migrating birds. Belgium has the highest number of SPAs under the Birds Directive (234) per country surface of all twenty-seven EU member states (the EU-27) (Ministerie van Economische Zaken, Landbouw en Innovatie, 2011).

The final responsibility for area designation rests in all member states at the ministerial level. In the federal states, formal designation can take place also under regional or provincial legislation (e.g. Belgium). The member states differ in the way they designate sites. Van Apeldoorn *et al.* (2010) discern two groups:

1 Those that favour general designation with only few site characteristics (name, location, borders and area; also species and habitats). Site objectives are worked out in detail in management plans. France and the UK are examples of this way of working.

2 Those that developed a detailed designation specifying site objectives and descriptions. This way of designation can be found in, among other countries, the Netherlands.

A general designation document may lead to uncertainty about the site objectives and hence more time consuming preparation of the management plan. A more detailed designation document, on the other hand, may lead to less time-consuming preparation of the management plan, but requires more site-specific information and thus takes more time to prepare.

The responsible ministries can decentralize some Natura 2000 implementation tasks to regional bodies, national committees, working groups or independent administrative bodies. A comparative study by Neven *et al.* (2005) found that functional decentralization in the Anglo-Saxon and Nordic culture leads to a more effective implementation process of Natura 2000. In the UK, Natural England is an intermediary depoliticized organization between central and decentralized levels of government in the specific area of nature conservation. This intermediary organization has the advantage of facilitating and mediating between policymakers and the policy area, which creates support among the stakeholders that implement the policy. Among the disadvantages are that it leads to fewer chances for the administration to integrate other priorities and that it diminishes political control. In France, mediation and negotiation takes the form of 'client' or 'relation' management, which is done at the local level by a prefect. In the Netherlands, decentralization is territorial and has been expressed through delegating certain tasks and authority for nature conservation (e.g. preparation of management plans) to the provinces. The management of many Natura 2000 sites is delegated to large environmental NGOs.

Both directives leave the issues of public consultation and the participation of stakeholders (such as experts, NGOs, lay public members and sector organizations) in site selection and designation to the member states. The member states largely differ in the way that they have dealt with these issues. For EU-27, Van Apeldoorn *et al.* (2010) categorize three types of participation:

1 Informal consultation: The public administration body produces information. The public has to find it and look for possibilities to express their opinion. Sometimes consultation meetings have been organized, the results of which may have been taken into account by the administrative authority.
2 Formal consultation and approval: The public administration body produces information. Stakeholders have legal rights to express their views and these have to be considered by the administrative authority.

3 Shared responsibilities (full participation): Stakeholders are organized and invited to express their opinions and are highly committed to and involved in decision making.

In the first phase of site selection and designation, 50 per cent of the EU-27 member states introduced informal participation. No member state organized participation in the form of shared responsibilities during the designation process. Even though NGOs were strongly involved by helping and advising the responsible ministry in many countries, this always occurred informally. In the later phase of designation most member states have opted for a more flexible strategy with at least some commitment. For instance, the responsible authorities in Finland and France shifted from little or no public participation to a more open way of working (Van Apeldoorn *et al.*, 2010).

Neven *et al.* (2005) have reported on the role of stakeholders in Natura 2000 implementation. According to this study, the implementation in the UK and France is characterized by goal-oriented communication with sector organizations combined with a bottom-up governance practice. Government provides a coaching style of leadership, facilitating contributions of public and private parties and 'learning by doing' strategies. Spain, France and Sweden apply similar bottom-up practices. In the Netherlands, the government exhibits an active and distinctly prescriptive attitude. The Ministry of Agriculture, Nature and Food Quality (now the Ministry of Economic Affairs, Agriculture and Innovation) outline memorandum for Natura 2000 published in 2005 indicates the most important steps in the implementation process and shows 'what must be done' and where there is scope for further consideration. In contrast to France and the UK, where freedom is given to bottom-up consultations with stakeholders, the Ministry of Agriculture, Nature and Food Quality has chosen a more indirect form of participation. In the Netherlands, information from the rounds of talks has been elaborated upon and processed by specialists and site managers into the 'Document for Natura 2000 Objectives' (Natura 2000 Doelendocument, cited in Neven *et al.*, 2005), which is used as a framework for implementation. Still, the relationship between government and environmental organizations is all too frequently characterized by a traditional top-down attitude and relationship (government instead of governance) with too little acknowledgement that the two need each other's resources and each other's knowledge. Other studies done in the Netherlands often stressed the importance of openness and flexibility in steering nature conservation policy (Natuurplanbureau, 1997; RMNO, 2003; RLG, 2003, all cited in Neven *et al.*, 2005, p. 112).

Similarly, Bouwma *et al.* (2008) discern two types of participation with regard to stakeholder involvement in the designation process in the neighbouring areas of the Netherlands: selective and professional participation. England, Lower Saxony and North-Rhine Westphalia carried out consultations at a local level with managers and owners of the sites (selective participation). By establishing working groups, England, Lower Saxony and North-Rhine Westphalia undertook an extensive site-based consultation process and an official procedure was established at the local level to enable complaints to be lodged prior to the list being sent to the EC. The Netherlands and Flanders involved only professional organizations (professional participation). Their contribution involved giving advice that was then taken into consideration in the policy process. The Netherlands and Flanders carried out consultations at the national and regional level without establishing an extensive site-based consultation process. The main targets included the authorities responsible for the areas and national organizations representing affected stakeholders.

For the neighbouring region of the Netherlands Bouwma *et al.* (2008) conclude that the selective participation in the designation process gave more opportunities to find the best solutions for all and thus generated stakeholder support and increased the legitimacy of later decisions. It also decreased the size of designated areas. Professional participation, on the other hand, deferred potential conflicts until the phase in which management plans were discussed with site owners. For the member states in general, the lack of participation of stakeholder groups in the designation process is the most common source of implementation problems (Paavola, 2004).

Management of Natura 2000 sites

The next step after formal designation is the establishment of necessary conservation measures to achieve the favourable conservation status at site level (in SPA and SAC). Both conservation measures and favourable conservation status need to be defined first, and a management plan is an instrument suggested by the Habitats Directive for this purpose. Only a few member states or regions, such as Flanders and the Netherlands, have formulated objectives at the national or regional level, which would be detailed at the site level. However, in many countries both objectives and plans are yet in their infancy (Kruk *et al.*, 2010).

The first assessment of conservation status based on the best available data covered the period 2001–2006. In the majority of member states, less than 40 per cent of species and habitats had a favourable conservation status (CBS, 2006).

Although management plans for Natura 2000 sites are only suggested by the EU legislation as an instrument to reach a favourable conservation status, they are obligated under national law or a site designation document in about a half of the EU-27 member states. In the remaining half, the plans are often written anyway as guidance for most of the sites. The majority of obligated management plans are also legally binding (Van Apeldoorn *et al.*, 2010; Kruk *et al.*, 2010). The Netherlands is alone among its neighbouring countries (England, Lower Saxony, North-Rhine Westphalia and Flanders) in that the formulation of management plans is obligatory by national law (Bouwma *et al.*, 2008). An important condition for achieving full implementation of management plans is obtaining high-acceptance levels by stakeholders who are involved in site management, especially where the management plans are not legally binding. In many cases, early and active stakeholder involvement in the designation process has contributed to their acceptance of the plans and their positive attitude towards Natura 2000. Case study results have shown that in a situation where management of the site depends on many different stakeholders, a planning process that emphasizes interaction and cooperation (the UK case) seems to offer better possibilities than a formal and hierarchical planning process (the Dutch case) (Beunen and de Vries, 2011). In Flanders, the original intention was to make management plans compulsory for all Natura 2000 sites. However, after drawing up six pilot schemes, Flanders decided to formulate management plans as an optional strategy. According to the Flemish environmental sector, the consensus-based approach has led to disappointingly low goal targets. The Flemish government agency stated that conducting these processes at every site was too time-consuming, and that the plan should only be used as a tool where suitable. However, Flemish farmers unions were positive about the value of a consensus-based approach to management plans (Bouwma *et al.*, 2008).

In most EU-27 member states, a management plan cannot force land owners and land users to take the necessary management measures, unless they have agreed voluntarily by signing a contract. The instrument of contract-based management is seen more and more as the basis for good site management in many EU countries. This is illustrated by the agri-environmental contracts that are used in all member states. Other instruments include letters of intent, licensing, integration of nature conservation into other programmes (e.g. rural and regional development), lists of permitted and forbidden activities, buying of land from private owners and execution of management (Kruk *et al.*, 2010). However, some issues remain with the financial compensation for land-use restrictions resulting from Natura 2000 and the extent to which interested individuals can participate in voluntary contract-based management. Another issue of concern is the general unwillingness of private landowners to join management schemes. However, in Flanders, the increased possibility of joining voluntary management schemes was viewed positively by land owners and land users, who had previously hostile emotions towards the imposed restrictions (Bouwma *et al.*, 2008).

Assessment of plans and projects under Article 6 of the Habitats Directive

Assessment of plans and projects according to Article 6 paragraphs 3 and 4 of the Habitats Directive, the so-called ‘habitat assessment’, constitutes the main instrument of Natura 2000. The assessment obliges the authorities to check if a plan or project is likely to have significant effects on a Natura 2000 site, and if that is the case, carry out an appropriate assessment of these effects (Article 6 paragraph 3). In case of a negative assessment, the authorities must check for possible alternatives; and if there are none, state the imperative reasons of overriding public interest and take compensation measures (Article 6 paragraph 4). Initially this formulation sparked considerable criticism due to undefined terms like 'significant effect', 'appropriate assessment', and 'overriding public interest'. These terms were later clarified through case law and guidance documents (European Commission, 2000, 2002, 2007a, 2007b).

The application and interpretation of Article 6 by the courts in England, Flanders, Austria, Germany, Sweden, France and the Netherlands has been investigated by Backes *et al.* (2006). On the whole, Article 6 was transposed reasonably well in most countries, although this has taken a considerable period of time and has caused friction. In the Netherlands, the courts have compelled direct application of Article 6, which caused a lot of problems in practice. In France, on the other hand, it seemed that projects that should have been tested against Article 6 have long profited from a tardy transposition because the rulings of local courts applying Article 6 were rejected by supreme courts. No direct check against Article 6 was encountered in England. This may be related to the fact that England had adopted the necessary implementation legislation as early as 1994.

Examples of differences in the transposition of Article 6 of the Habitats Directive in the Netherlands and Flanders are given by Van Hoorick (2004) and Freriks (2004). Article 6 of the Habitats Directive was transposed in Article 36 paragraphs 3–7 of the Nature Conservation Decree in Flanders in 2002. Appropriate assessment in Flanders applies to potentially adverse activities requiring a permit, plans or programmes. The notions 'activity requiring a permit' as well as 'plan or programme' are defined in the opening of the Nature Conservation Decree. Not all countries provide a statutory definition of a 'plan' in their national law (Flanders and Germany are the only ones among those studied by Backes *et al.* 2006). Flemish transposition typically uses 'activity requiring a permit, plan or programme' instead of 'plan or project' as in the Habitats Directive. The consequence is that certain activities that do not require a permit under the national permit system fall outside the scope of appropriate assessment. It is a shortcoming in the Flemish region that there is no separate permit requirement for these activities (Van Hoorick, 2004). The requirements for compensating measures (Article 36 paragraph 5-2) as stated in the Nature Conservation Decree follow the interpretation guide of the European Commission quite closely, and are perhaps stricter than the original text of Article 6 paragraph 4 Habitats Directive. They require a compensation measure that actively develops an equal habitat or an equivalent surface area of its natural environment.

Article 6 of the Habitats Directive was transposed in Articles 19d–19k of the Nature Protection Act in the Netherlands in 2005. The Dutch Nature Protection Act links to the definitions from the Directive (plan/project) and adds the category 'other activities'. These categories are broader than the ones found in the Flemish legislation. The decisions about 'habitat assessment' in the Netherlands are issued by the provincial authorities as a certificate of incorporation. The Flemish legislation stipulates that in a specific case – the case of overriding public interest – it is always the Flemish government that decides on the existence of overriding public interest, including social and economic considerations (Article 36 paragraph 5). The consequences for a project in the case that the Flemish government does not follow up on the reasons of overriding public interest are therefore not clear (Freriks, 2004).

Furthermore, Backes *et al.* (2006) found that there is a relationship between the roles the courts either do, or do not, play and the meaning of Article 6 in practice. This is explicitly linked to restrictions in access to the courts and the different ways justice is administered within the EU. In both Austria and Germany, there are very limited possibilities for public participation and legal protection available to citizens to contest the national legislation that puts Article 6 paragraphs 3 and 4 of the Habitats Directive into effect. The highest court considers that legal protection of citizens is not required since EU nature conservation directives concern the protection of a general interest as opposed to individual interest. This is not regarded by the European Commission as an infringement of EU law, as citizen involvement is not mandatory in Article 6 paragraph 3. Public participation is mandatory if there is also an environmental impact assessment procedure, but this derives from the Environmental Impact Assessment Directive and not from the Habitats Directive. In England, it is decided in each specific case whether opportunities for public participation are offered and legal protection is often available against sector decisions. There is relatively less haste in England to present differences of opinion on the application of legal frameworks to the courts, and solutions are first sought through consultations. In Sweden, the way the general public is consulted depends on the statute (e.g. physical planning law, etc.) that the decision was taken upon and is usually limited to the 'affected public' and environmental NGOs. In Flanders, environmental NGOs have access to the Belgian Council of State to contest decisions. In the Netherlands, access to the appeals procedure against a permit issued under the Nature Protection Act is open to interested parties and stakeholders including environmental NGOs. In the Netherlands, unlike in Germany and Austria, there is no relativity review: an interested party can present all arguments against a decision no matter whether he or she has a specific interest with these arguments. Opportunities for participation in legal actions in the Netherlands are thus much wider compared to the other countries studied.

Another finding by Backes *et al.* (2006) is related to the role of information and guidance as a facilitating measure by the government for proper application of Article 6. In the countries studied, the authorities responsible for transposing and implementing Article 6 published guides or explanatory memorandums on the interpretation of Article 6 earlier than in the Netherlands. The need for information 'assistance' is particularly high in France, Germany and England, where there is no separate permit for nature conservation, but the decisions on application of Article 6 paragraphs 3 and 4 are taken within the framework of other permit systems directed more at the approval of plans and projects. The issuing authority in these countries is very often not primarily an expert in the field of nature conservation and is very often advised by an expert government agency, like Natural England in England. The study notes that an expert centrally organized authority like Natural England can prove particularly beneficial for the quality of application of Article 6 Habitats Directive.

14.3 The 'Building with Nature' Approach

The concept of 'Building with Nature' was introduced by the Czech hydraulic engineer Svašek in 1979 and was further explored and linked to the field of integrated coastal management by Waterman (2008, 2010). Waterman defines 'Building with Nature' as flexible integration of coast and water by making use of materials, forces and interactions present in nature, in the context of the hydrological and morphological situation (Waterman, 2010). Sector organizations and supranational authorities have also adopted the concept, although under slightly different names. The process of adoption emerged naturally, as the pressure on estuaries and coasts is experienced by multiple actors due increasing economic activities in these areas. The World Association for Waterborne Transport Infrastructure (PIANC) published its first position paper 'Working with Nature' in 2008 and revised it in 2011. PIANC sees 'Working with Nature' as doing things in a different order: establish project needs and objectives, understand the environment, make meaningful use of stakeholder engagement and prepare initial project design to benefit navigation and nature (PIANC, 2011). In 2011 the European Commission adopted specific guidelines for dealing with the Birds and Habitats Directives in estuaries and coastal zones (European Commission, 2011). This document stresses that project design 'should always be based on mutually beneficial strategies with a view to achieving dual goals of both Natura 2000 conservation objectives and socio-economic objectives, according to the "working with nature" concept' (European Commission, 2011, p. 5). The authorities and practitioners hope that by integrating environmental concerns as early as possible in the project design stage, they serve the environment and society better and improve project implementation in the field of coastal and delta infrastructure.

In the Netherlands, the EcoShape Foundation is at the moment the main proponent of this approach (EcoShape, 2012). The foundation applies the Building with Nature approach to large-scale sand nourishments and ecological landscaping in the Netherlands by adopting three guiding principles: make optimal use of natural forces and processes present in the ecosystem; explore opportunities for nature development as an integral component of project design and reserve space to accommodate for natural system dynamics (Aarninkhof *et al.*, 2010). Although it is too early for the best practices of the approach to be established, more and more projects are beginning to show one or several features of this new approach:

- Exploring opportunities for nature development at the initial project design stage and integrating socio-economic (e.g. flood defence, recreation) and ecological project goals.
- Using nature dynamics and materials occurring in nature in the context of hydrological and morphological situation to achieve the project's goals.
- Creating opportunities for development of new nature, improving the ecological values currently present in the project area.

Building with Nature can thus be defined as an approach that uses natural dynamics to achieve project objectives and integrates ecological value into the outcome. The approach gives natural dynamics a prominent place, while the goal of nature conservation policy is to conserve the existing values. Hence nature conservation policy is an important contextual element related to Building with Nature realization and transfer. Initially, Building with Nature proponents expected that the conservation goals of Natura 2000 would block the efforts to develop ecosystems. However, empirical evidence has shown that this is not necessarily the case (Vikolainen *et al.*, 2011, 2012). More interesting is that Building with Nature – under the name of Working with Nature – has the potential to become a water management 'best practice' in Natura 2000 areas. The European Commission encourages its application for port development, dredging and inland waterway development and management in the context of the EU Birds and Habitats Directives (European Commission, 2011, 2012). In the next section we discuss the underlying dynamics of this process.

14.4 Discussion

The first research question asked in this chapter was: what does the transfer of Natura 2000 in north-west EU member states look like? From the literature reviewed in this chapter it became clear that the transfer of the Natura 2000 policy from the supranational level to the member-state level varies across member states. Nonetheless, it is possible to make a conceptual representation of the factors that constitute the Natura 2000 context. The

Natura 2000 factors discussed in Section 14.2 can be divided into two categories: site-related and governance-related. Using the insights from CIT, these factors are grouped according to the contextual layers as specific and structural context factors as shown in Figure 14.1. The member states differ in their choices of legal framework for the transposition of the EU directives and especially Article 6 of the Habitats Directive, the participation arrangements for stakeholders affected by designation, the possibilities for the general public to contest the decisions regarding implementation of the directives' provisions, the availability of national information and guidance for the directives' implementation and the authorities responsible for national implementation. Additionally, each member state defines the size and borders of Natura 2000 sites, determines whether management plans and/or contracts will be used, defines the conservation status and sets the conservation objectives for the site.

Prospective transfer evaluation requires information about programmes that are in effect elsewhere as well as awareness of the differences in policy settings (Mossberger and Wolman, 2003). This is often problematic, especially in complex policy arenas like that of Natura 2000 with its many variables to consider. The representation of the Natura 2000 context shown in Figure 14.1 could serve the purpose of creating awareness about the Natura 2000 context for the prospective transfer of the Building with Nature approach within the EU. The next step would be to isolate the factors that matter and gain sufficient knowledge for understanding the impact of those factors, which is not an easy task. Figure 14.1 suggests that specific context factors, which include Natura 2000 site size, borders, available management

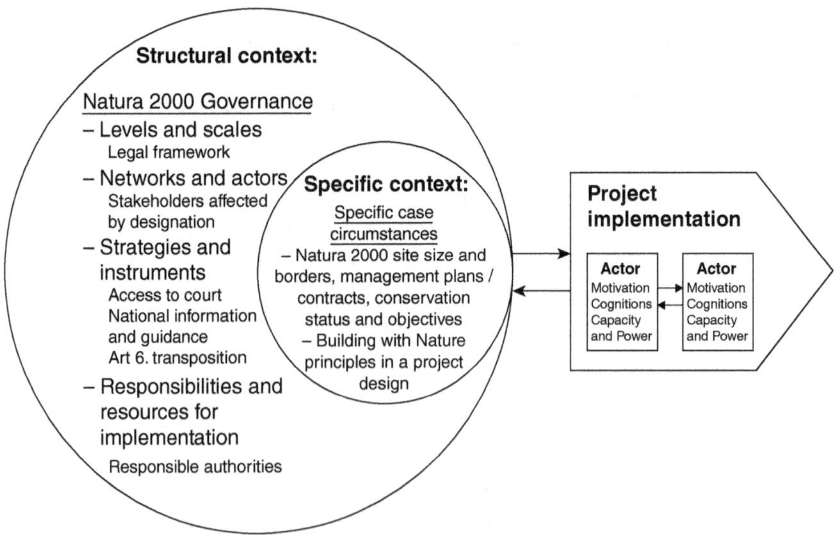

Figure 14.1 The transfer of Natura 2000 to the member-state level.

plans and/or contracts, conservation status and objectives, constitute the immediate environment where a Building with Nature project would take place. As Mossberger and Wolman (2003) suggest, prospective evaluation hinges on the ability to predict the adaptations required for successful implementation. If any adaptations were to be made to a Building with Nature project to increase its effectiveness within the Natura 2000 context, they should be related to the specific context factors of Natura 2000. For example, empirical evidence has shown that a project design that explicitly takes into consideration the conservation objectives of Natura 2000 site has more chances of success if an appeal is made to an administrative court (Vikolainen *et al.*, 2011). Another possibility that deserves further investigation is the introduction of Building with Nature principles in Natura 2000 management plans, which are currently in preparation in several member states. Specific context, in turn, should be considered against the background of structural context factors, which vary across member states.

The second research question asked in this chapter was: what are the opportunities for successful transfer of Building with Nature approach within the EU Natura 2000 context? The application of Building with Nature is encouraged by the European Commission for port development, dredging and inland waterway development and management in the context of the EU Birds and Habitats Directives (European Commission, 2011, 2012). The concept of best practice (or good practice) is rife in European policies and programmes (Stead, 2012). The strategy of encouraging best practices and models of behaviour pursued by the European Commission is efficient for building 'technocratic' legitimacy: the Commission cannot be accused of trying to impose 'the view of Brussels' if it follows a policy design that is already in place somewhere in the EU (Radaelli, 2000). However, initial expectations of Building with Nature proponents were quite different, if not the opposite. In particular the private members of the EcoShape consortium expected the efforts of eco-dynamic design and development to be blocked by the Natura 2000 conservation objectives, which they perceived as too rigid.

Although Building with Nature is not yet fully in place in the Netherlands, it is taking root in Dutch water management. In 2008, the commission appointed by the Dutch government to address the long-term threats of climate change recommended the application of Building with Nature principles for beach and shoreface nourishments, which include the primary measures to guarantee long-term safety and development of coastal zones (Delta Commission, 2008). According to Tews *et al.* (2003) 'high' environmental standards of pioneering nation-states often find their way into international agreements and recommendations with weak enforcement mechanisms. From this point of view, it is not surprising that innovative approaches like Building or Working with Nature are recommended in the European Commission Guidelines. Such 'aspirational' arrangements usually reflect the agenda-setting power of ambitious, well-organized private actors from 'forerunner' countries (e.g. EcoShape). At the same time, transnational non-state actors facilitate national adoptions of policy innovations practised in other countries or modelled on internationally promoted 'best practices'. This is what the World Association for Waterborne Transport Infrastructure (PIANC) does in its position paper 'Working with Nature' (PIANC, 2011). Once new approaches in environmental policy have been put into practice in 'forerunner' countries, it becomes increasingly difficult for nation-states to resist adopting the same approach without threatening their image as legitimate members of environmentally responsible global society (Tews *et al.*, 2003). The adoption of Building with Nature is also favoured by the underlying problem structure: unwanted outcomes associated with water infrastructure development in Natura 2000 estuaries and coasts – in the form of project delays and cancellations – are directly visible and widely discussed by politicians and general public.

For the reasons discussed above, Building with Nature stands a considerable chance of becoming a 'best practice' for water management in estuaries and coastal zones. The underlying assumption is that best practices are equally applicable and effective in another setting. However, as pointed out by Stead (2012), best practices are unlikely to lead to the same outcomes across different EU member states no matter how faithfully they are transferred. The applicability and effectiveness of Building with Nature transfers could increase if they take into account the Natura 2000 context. The framework offered in this chapter could be useful for doing this.

14.5 Conclusion

This chapter focused on two types of transfers within the EU: the coercive transfer of the Natura 2000 biodiversity policy and the prospective transfer of Building with Nature. Although the transfer of Natura 2000 varies across member states, the core differences can be described along the nine governance- and site-related contextual factors depicted in Figure 14.1. For policymakers evaluating the prospective transfer of Building with Nature, this representation of the Natura 2000 context could be useful for becoming aware of and adapting their transfer efforts to the Natura 2000 context in different member states. The transfer of Building with Nature is likely to be favoured by the dynamics behind the adoption of environmental 'best practices' introduced by 'forerunner' countries as well as the underlying problem structure, but its applicability and effectiveness could increase if it takes into account the Natura 2000 context.

Acknowledgement

The work presented in this chapter was carried out as part of the innovation programme Building with Nature. The Building with Nature programme is funded from several sources including the Subsidieregeling Innovatieketen Water (SIW, Staatscourant nr. 953 and 17009), sponsored by the Dutch Ministry of Transport, Public Works and Water Management, and partner contributions from the participants in the Foundation EcoShape. The programme receives co-funding from the European Fund for Regional Development (EFRO) and the Municipality of Dordrecht.

References

Aarninkhof, S.G.J., van Dalfsen, J.A., Mulder, J.P.M. and Rijks, D. (2010) *Sustainable Development of Nourished Shorelines. Innovations in Project Design and Realisation,* paper presented at PIANC MMX Congress, Liverpool.

Arnouts, R.C.M. and Kistenkas, F.H. (2011) *Nederland op slot door Natura 2000: de discussie ontrafeld [The Netherlands locked up by Natura 2000: the discussion unravelled],* Werkdocument 236. Wettelijke Onderzoekstaken Natuur & Milieu, Alterra, Wageningen.

Backes, C.W., Freriks, A.A. and Nijmeijer, A.G.A. (2006) *Article 6 Habitats Directive: A comparative law study on the implementation of Article 6 Habitats Directive in some member states,* Centrum voor Omgevingsrecht en Beleid/NILOS, Utrecht.

Bastmeijer, C.J. and Verschuuren, J.M. (2003) *Knelpunten bij de Uitvoering van de Natuurbeschermingswetgeving in Nederland [Bottlenecks in the implementation of Nature Conservation Legislation in the Netherlands],* Eindrapport in opdracht van Ministerie van Financiën. Centrum voor Wetgevingsvraagstukken, Universiteit van Tilburg, Tilburg.

Bastmeijer, C.J. and Verschuuren, J.M. (2004) 'Juridische Knelpunten bij de Bescherming van Natuurgebieden' [Legislative bottlenecks in the protection of nature sites], *Milieu en Recht,* vol. 31, no. 3, pp. 142–149.

Beijen, B.A (2010) 'De Kwaliteit van Milieurichtlijnen. Europese Wetgeving als Oorzaak van Implementatieproblemen' [The quality of Environmental Directives. European legislation as a source of implementation problems], PhD thesis, University of Utrecht.

Beunen, R. (2006) 'European nature conservation legislation and spatial planning: For better or for worse?' *Journal of Environmental Planning and Management,* vol. 49, no. 4, pp. 605–619.

Beunen, R. and de Vries, J.R. (2011) 'The governance of Natura 2000 sites: The importance of initial choices in the organization of planning processes', *Journal of Environmental Planning and Management,* vol. 54, no. 8, pp. 1041–1059.

Beunen, R., van der Knaap, W.G.M. and Biesbroek, G.R. (2009) 'Implementation and integration of EU environmental directives: Experiences from the Netherlands', *Environmental Policy and Governance,* vol. 19, no. 1, pp. 57–69.

Bouwma, I.M., Kamphorst, D.A., Beunen, R. and van Apeldoorn, R.C. (2008) *Natura 2000 Benchmark: A comparative analysis of the discussion on Natura 2000 management issues,* Rapport 90, Wettelijke Onderzoekstaken Natuur & Milieu, Alterra, Wageningen.

CBS (2006) Centraal Bureau voor de Statistiek. Staat van instandhouding Habitats en Soorten, online, available at: www.compendiumvoordeleefomgeving.nl, accessed 11 November 2011.

Delta Commission (2008) *Working Together With Water: A living land builds for its future – Findings of the Deltacommissie 2008, Summary and conclusions,* Delta Commission: the Hague, online, available at: www.deltacommissie.com/doc/deltareport_summary.pdf, accessed 2 August 2012.

Dolowitz, D.P. and Marsh, D. (1996) 'Who learns what from whom: A review of the policy transfer literature', *Political Studies,* vol. 44, no. 2, pp. 343–357.

EcoShape (2012) 'Building with Nature', www.ecoshape.nl, accessed 12 March 2012.

EEC (1979) Council Directive 79/409/EEC of 25 April 1979 on the Conservation of Wild Birds, *Official Journal L 103,* pp. 0001–0018.

EEC (1992) Council Directive 92/43/EEC of 22 July 1992 on the Conservation of Natural Habitats and of Wild Fauna and Flora, *Official Journal L 206,* pp. 0007–0050.

European Commission (2000) *Managing Natura 2000 Sites: The provisions of Article 6 of the Habitats Directive 92/43/EEC,* Office for Official Publications of the European Communities, Luxembourg.

European Commission (2002) *Assessment of Plans and Projects Significantly Affecting Natura 2000 Sites: Methodological guidance on the provisions of Articles 6(3) and (4) of the Habitats Directive 92/43/EEC,* Office for Official Publications of the European Communities, Luxembourg.

European Commission (2007a) *Guidance Document on Article 6(4) of the 'Habitats Directive' 92/43/EEC: Clarification of the concepts of: alternative solutions, imperative reasons of overriding public interest, compensatory measures, overall coherence, opinion of the Commission,* Office for Official Publications of the European Communities, Luxembourg.

European Commission (2007b) *Guidance Document on the Strict Protection of Animal Species of Community Interest under the Habitats Directive 92/43/EEC,* Office for Official Publications of the European Communities, Luxembourg.

European Commission (2011) *Guidelines on the Implementation of the Birds and Habitats Directives in Estuaries and Coastal Zones, with Particular Attention to Port Development and Dredging,* Office for Official Publications of the European Communities, Luxembourg.

European Commission (2012) *Guidance on Sustainable Inland Waterway Development and Management in the context of the EU Birds and Habitats Directives,* Office for Official Publications of the European Communities, Luxembourg.

Freriks, A. (2004) 'Gebiedsbescherming op grond van de Vogelrichtlijn en de Habitatrichtlijn: implementatie in Nederland [Site protection on the basis of the Birds and Habitats Directives: implementation in the Netherlands]', in F.C.M.A. Michiels and L. Lavrysen (eds) *Milieurecht in de lage landen,* Boom Juridische Uitgevers, Den Haag.

Kruk, R.W., de Blust, G., Van Apeldoorn, R.C., Bouwma, I.M. and Sier, A.R.J. (2010) *Natura 2000: Information and communication on the designation and management of Natura 2000 sites,* Summary Main report 2: Organizing the management in 27 EU member states. Alterra-rapport 2044.1./INBO.R 2010.28, Alterra, Wageningen.

Ministerie van Economische Zaken, Landbouw en Innovatie (2011) *Implementatie Natura 2000 in Nederland. Analyse naar aanleiding van het Regeerakkoord en de motie van der Staaij c.s. naar nationale koppen, rek en ruimte in de Natura 2000-implementatie*, Den Haag: Ministerie van Economische Zaken, Landbouw en Innovatie.

Mink, F.J. (2007) *Maritime Infrastructure and Marine Coastal Zones: Issues with the Habitats Directives*, European Dredging Association Paper 06-0347, European Dredging Association, Brussels.

Mossberger, K. and Wolman, H. (2003) 'Policy transfer as a form of prospective policy evaluation: Challenges and recommendations', *Public Administration Review*, vol. 63, no. 4, pp. 428–440.

Natura 2000 Barometer (2010) online, available at: www.ec.europa.eu/environment/nature/natura2000/barometer/index_en.htm, accessed 11 November 2011.

Neven, M.G.G., Kistenkas, F.H. and van Apeldoorn, R.C. (2005) *Eurosites Insights: Image, implementation, interpretation and integration of Natura 2000 in European perspective*, Alterra-rapport 12222.1, Wageningen.

Paavola, J. (2004) 'Protected areas governance and justice: Theory and the European Union's Habitats Directive', *Environmental Sciences*, vol. 1, no. 1, pp. 59–77.

PIANC (The World Association for Waterborne Transport Infrastructure) (2011) PIANC Position Paper 'Working with Nature', www.pianc.org/workingwithnature.php, accessed 14 March 2012.

Radaelli, C.M. (2000) 'Policy transfer in the European Union: Institutional isomorphism as a source of legitimacy', *Governance*, vol. 13, no. 1, pp. 25–43.

Stead, D. (2012) 'Best practices and policy transfer in spatial planning', *Planning Practice and Research*, vol. 27, no. 1, pp. 103–116.

Tews, K., Busch, P. and Jörgens, H. (2003) 'The diffusion of new environmental policy instruments', *European Journal of Political Research*, vol. 42, no. 4, pp. 569–600.

Van Apeldoorn, R.C., Kruk, R.W., Bouwma, I.M., Ferranti, F., de Blust, G. and Sier, A.R.J. (2010) *Natura 2000: Information and communication on the designation and management of Natura 2000 sites*, Summary Main report 1: the designation in 27 EU member states, Alterra-rapport 2043.1./INBO.R 2010.24, Alterra, Wageningen.

van der Zouwen, M. and van Tatenhove, J.P.M. (2002) *Implementatie van Europees natuurbeleid in Nederland [The Implementation of European Nature Policy in the Netherlands]*, Planbureaustudies nr. 1, Natuurplanbureau, Wageningen.

Van Hoorick, G. (2004) 'De implementatie van "Natura 2000" in het Vlaamse Gewest [The implementation of Natura 2000 in the Flemish region]', in F.C.M.A. Michiels and L. Lavrysen (eds) *Milieurecht in de lage landen*, Boom Juridische Uitgevers, Den Haag.

Van Hooydonk, E. (2006) *The Impact of EU Environmental Law on Waterways and Ports*, Sixth Framework Programme Sustainable Surface Transport document M22.00.24.052.001, Maklu Press, Antwerp.

Verschuuren, J.M. (2003) 'Nogmaals kokkels, hamsters en woelmuizen: recente ontwikkelingen rond de implementatie van de Vogel- en Habitatrichtlijn in Nederland' [Once again cockles, hamsters and voles: Recent developments concerning the implementation of the Birds and Habitats Directives in the Netherlands], *Nederlands Tijdschrift voor Europees Recht*, vol. 9, no. 1/2, pp. 34–38.

Verschuuren, J.M. (2004) 'Implementatie milieurichtlijnen: Nederland raakt achterop' [The implementation of environmental directives: The Netherlands is lagging behind], *Nederlands Tijdschrift voor Europees Recht*, vol. 10, no. 12, pp. 348–351.

Vikolainen, V., Bressers, J.T.A. and Lulofs, K.R.D. (2011) 'Management aspects of Building with Nature projects in the context of EU Birds and Habitats Directives', *Environmental Engineering and Management Journal*, vol. 10, no. 11, pp. 1675–1686.

Vikolainen, V., Bressers, J.T.A. and Lulofs, K.R.D. (2012) 'Implementing EU Natura 2000 at the project level: Lessons from the Veluwe Border lakes in the Netherlands', *Public Administration*, vol. 90, no. 4, pp. 1016–1031.

Waterman, R.E. (2008) *Towards an Integrated Coastal Policy via Building with Nature*, Waterman, Delft.

Waterman, R.E. (2010) 'Integrated coastal policy via Building with Nature', PhD thesis, University of Delft, the Netherlands.

15 Conclusions

Joanne Vinke-de Kruijf, Gül Özerol, Cheryl de Boer and Hans Bressers

15.1 Introduction

The chapters of this book have highlighted the many pathways through which concepts related to water governance are emerging and being transferred in countries all around the world. The motivating purpose of collecting these chapters into one volume was to begin a critically-based understanding of the balance between learning and implementing lessons from abroad, and recognizing the special character of local contexts when assessing appropriate water governance approaches at a given scale (local, regional, national or international). We use this concluding chapter to accentuate the interesting aspects of these transfers, to make some notes on the commonalities that have been found, and to discuss what this means for possibilities to generalize statements about the transferability of water management knowledge. These reflections are made on the basis of relevant theoretical insights (presented in Chapter 2 and Chapter 3) and empirical findings (see Table 1.1 in Chapter 1 for an overview of Chapters 4 to 14).

Chapter 2 begins with an introduction of some of the concepts that are widely promoted by international organizations as solutions to different water problems being experienced around the world. A number of empirical chapters present concrete experiences with the translation of these concepts. For example, they discuss how 'integrated water resources management' is formalized in Kazakhstan or how 'public participation' is given meaning in Turkey's water management system. Chapter 2 also introduces the phenomenon of 'policy transfer' as an interactive process that may involve various types of knowledge, a wide range of actors and result from various factors. Transfers in the water management sector can take different forms as the diversity of the empirical cases reflects. The cases illustrate that questions about the transferability of knowledge arise when projects are initiated for the purpose of transferring concrete knowledge from one country to another (Chapters 4 and 5) or when international commitments or principles get implemented (or not) on the ground (Chapters 7 to 11). Such questions also arise when reflecting upon existing or emerging 'best practices' (Chapters 6 and 14) or upon governance regimes' need to adapt (Chapters 12 and 13). On the basis of the literature on 'transfers', Chapter 2 also presents some ideas about what distinguishes successful from less successful transfers and how the transferability of knowledge is influenced by contextual factors. In Chapter 3, we elaborate a theoretical lens – the Contextual Interaction Theory – that is used in various chapters to understand policy transfers. In this theory, the process and outcomes of interactive processes are shaped by the interactions and characteristics of actors in relation to the context in which they occur. This includes the wider context as well as the governance context. Various authors use this theory to analyse both the characteristics of actors and the wider and governance contexts (in particular Chapters 4, 5 and 11). Others focus upon the theory's assertions about governance qualities (in particular Chapters 10 and 13). Furthermore, some authors build more directly upon policy transfer literature (Chapters 6 and 7).

In this concluding chapter, we reflect upon the broad range of experiences that are presented in various empirical chapters. First, we reflect on the empirical findings using three lines of understanding: the transfer of established policy knowledge (Chapters 4 to 6), the transfer of policy concepts that are part of global discourses (Chapters 7 to 11) and the emergence of policy concepts (Chapters 12 to 14). Following this, we reflect upon the methods used for studying transfers and discuss the possibilities for future studies on this topic. Section 15.3 reflects on the impacts these results have had on our understanding of the importance and appropriateness of contextual water management. Finally, we choose to conclude by reflecting on the value of critically discussing the feasibility of policy transfers for modern water governance.

15.2 Reflections on empirical findings

As one of the central questions of this book is related to understanding how, where and why context affects actors involved in a transfer process, we have chosen to make use of a broad range of experiences in the included chapters. This wide variety of cases provides us with examples of various pathways in which transfers of knowledge about water management and governance concepts take place. We organized the empirical chapters along three lines of understanding that were introduced in the introductory chapter (Chapter 1). This reflection on the empirical findings is organized accordingly. While discussing the findings of these three sections, we reflect upon the descriptive, explanatory and evaluative questions that are introduced in Chapter 1. This includes questions about the transfer process, the context(s) of the transfer under study and the effectiveness or benefits of the transfer.

Transfer of established knowledge

The transfer of knowledge that was developed and applied in one country and subsequently used in the development of policies or practices in another country is central in the first three empirical chapters of the book. The first two chapters describe cases that involve the actual transfer of knowledge from a developed country to a transition country, namely from the Netherlands to Romania (Chapter 4) and from the United States to India (Chapter 5). Chapter 6 presents a prospective evaluation by assessing the potential of such a transfer from a Canadian province to elsewhere in the world.

The chapters on the Romanian and the Indian cases both use insights of the Contextual Interaction Theory to understand the interactive process that evolved around the transfers under study. Regarding the actors involved, international actors were not directly involved in the Romanian and the Indian cases. In both cases, the transfers were initiated by external experts (such as private consultants, water managers or faculty–student teams) who implemented the project in collaboration with various stakeholders of the receiving country. The analyses of actor characteristics provide insights regarding the nature of the transfers. Both transfers were driven by a desire of the transferring actors to share their knowledge and characterized by an unequal distribution of resources. The authors of both chapters further observe that receiving actors were often not as convinced about the transfer as the transferring actors but became more motivated over the course of time. Therefore, the transfers are not considered to be imposed but also not as completely supported by the receiving actors. The willingness of the receiving actors to contribute increased as a result of their ongoing interactions. The Contextual Interaction Theory adds here to the literature on policy transfer by drawing attention to the diverse and dynamic nature of transfer processes. It sheds light on the diverse motivations of actors to engage in such processes and how these motivations are likely to change over time.

Regarding the knowledge being transferred, the respective chapters focus on knowledge about flood risk management, water drainage and drinking water source protection that had been previously applied. The 'universal remedies' that were introduced in Chapter 2 are not the objective of the transfer in any of the three chapters. However, all transfers were or would be implemented under the influence of a water governance panacea – integrated flood management in the Romanian case and collaborative water management in the Indian and Canadian cases. Thus, even transfers of established knowledge are not limited to 'specific' solutions but come along with broader ideas and practices.

What characterizes the 'context' in which solutions are developed is analysed in detail in the Canadian case. The authors distinguish between various indicators of transferability related to the policy problem, the characteristics of the policy being transferred and the policy context. They conclude that aspects of the water governance context – including prevailing participation and integration practices and the division of responsibilities – require special attention since they are closely interwoven with the knowledge being transferred. The Romanian case confirms this prospective evaluation of transferability. The authors conclude that the effectiveness – or rather the lack of effectiveness – of the cases is correlated with how well knowledge is embedded (or not) into Romania's governance context. The authors also analyse differences in the wider contexts of the transferring and the receiving countries and conclude that contextual differences do not only complicate the policy transfers, but also constitute an incentive for policy transfers.

The Romanian case and the Indian case use different methods to evaluate their respective transfers. The authors of the Romanian case use 'the degree to which the objectives of the transferring country are achieved' as the indicator of effectiveness. They conclude that none of the analysed transfers made a direct contribution to problem solving and that only one of the transfers is likely to result in direct follow-up actions that would benefit the Dutch water sector. Probable explanations are sought in contextual factors and appear to be closely related to the involvement and support of national actors. The authors of the chapter on the Indian case base their assessment on the criterion of 'true collaboration' (i.e. active cooperation rather than passive or forced cooperation), which is operationalized by five traits: purpose; dialogue; decision making; action; and evaluation. They identify areas of improvement in decision making and action, and conclude that this is related to the lack of a formal (collaborative) decision process and structure. Thus, the involvement and commitment of the receiving actors emerges as an important factor in realizing successful transfers. The crucial importance of the receiving actors also appears from the conclusions of Chapter 6: to realize the benefits of policy transfer, receiving actors should understand the factors that shape transferability and develop the capacity to utilize the lessons. Chapters 4 and 5 show that in cases of an 'unequal' transfer (e.g. from a developed country to a less developed country), transferring actors should ensure the involvement and commitment of the receiving actors. In this regard, the authors of Chapter 5 conclude that the Contextual Interaction Theory can help to recognize – and preferably to prevent – when coercion is taking place.

Transfer of international concepts

Following the idea that several policy concepts are being promoted by various international organizations, five chapters in this book focus on the transfer – or the translation – of these policy concepts into new contexts. These chapters include cases regarding the transfer of internationally promoted concepts to Kazakhstan's national water policy (Chapter 7), regional water management in Turkey (Chapter 8), water-related local projects in Palestine (Chapter 9), urban water management in Vietnam (Chapter 10) and river basin management in Mexico (Chapter 11).

Central to the analysed cases is the application of concepts that are closely linked to ideas about 'integration' and 'participation' that were presented in Chapter 2. In all chapters, international organizations played a role in the initiation of the transfers. It is therefore debateable to what extent these translations took place through entirely voluntary efforts from the receiving countries. Developing countries (including low- and middle-income countries and emerging economies) often feel pressure to adopt and implement 'first-world' solutions. This pressure can come in many forms and will be experienced differently based on the specific, structural and wider context of the case being studied. In these chapters the motivations to adopt new strategies were related to various desires:

1 the general desire to improve the governance of resources based on the belief that these new panaceas would help solve current issues;
2 the desire to be eligible for development aid; and
3 the desire to participate in international discourses as an equal partner.

In the case of the Kazakhstan chapter, the translation of the policy concepts into the national discourse and legal institutions are influenced by international development projects, discourses, actors and funding. In the Turkish case, the authors argue that the transition towards more participatory water management is characterized by a top-down imposition by successive governments and international agencies. The chapter on the Palestinian case examines multi-stakeholder partnerships (MSPs) in water and sanitation projects, and concludes that the MSPs emerge as an empirical practice with or without external pressure from donor organizations. In the Vietnamese case, the author analyses the transfer from the perspective of international organizations. He shows that integrated urban water management involves very complex and hierarchical governance structures. Therefore, international commitments referring to 'integration' cannot be transferred to the local level without an accompanying change in the overall culture and governance structure of Vietnam. In the Mexican case, the river basin councils were not only a response to international studies and influence of international agreements but also to urgent problem pressures and were further enabled by changes in the political and economic context.

Various chapters show that questions about the success or benefits of the transfers of international concepts are closely related to questions about the context to which knowledge is being translated. In the Mexican case, there is a lack of progress made in increasing the water quality in the focus areas. This, however, is only one of the indicators being used by the local participants in determining the successfulness of their actions. General ecosystem improvement was seen as an additionally relevant goal, one that was improved as a result of actions designed to improve water quality and prevent fish-kills. The studied river basin commission emerged in a bottom-up fashion and was thus not experienced as an external intrusion into the governance context. Issues did develop, however, with respect to the river basin commission's ability to be seen as directly connected to the new commitments for water management such as watershed-based governance and stakeholder participation. Constructive stakeholder participation led to a joint vision of the future of the basin and common understanding and joint ambitions developed over time. A lack of financial and legal support and minimum public attention to their contributions is proposed to have resulted in a governance regime context that is only in principle supportive. The authors conclude that the governance regime does not adequately equip organizations like the basin commission with sufficient support for its integrated ambitions for the basin. While river basin management and public participation are realized to a large extent and in a productive way in the commission itself, the integration of the commission's process into the context of the mainstream policy process remains challenging. In the Vietnamese case, the desire to improve water management cannot be decoupled from the more immediate goals of poverty alleviation and general economic development. Thus, while there is a lack of capacity to fully implement integrated water resource management in the manner which is envisioned by the many international visioning and policy documents, the improvements made in increasing regulatory control, providing green belts for urban areas and recognizing urban agriculture as a stakeholder in the city's development are considered important advances in addressing the more immediate issues related to their primary goals. In the case of Kazakhstan, the author describes the translation of policy innovations as a 'bumpy ride with unplanned consequences and great complexities'. As one of the major hurdles facing the implementation of these policies, the author mentions the lack of public deliberation options. He recognizes however that the development of capacity for policy deliberation is a long-term process, and one that is not common in post-Soviet regions. He concludes that the special context of Kazakhstan, in terms of its history of water and land management as well as the characteristics of water policy and politics (such as the presence or absence of a forum for policy deliberation), have contributed to the way that these policy concepts have been translated. Regarding the role of contextual factors, similar conclusions are reached in the Turkish case. Despite the strong imposition by international organizations, contextual factors (mainly the pre-existing hierarchical structure and the related lack of experience with participatory mechanisms) limit the success of transfer by leading to the contestation of public participation. The authors also pinpoint the role that actor characteristics (mainly power asymmetries and insufficient local capacities) play in the contested nature of public participation. In the Palestine case, the authors reveal that the multiplicity in the fragile structural context (actors, structures, regulations and practices) facilitates the emergence of MSPs as a practice, while at the same time hampering the effectiveness and complicating the governance and cooperation processes. The authors attribute these limitations in implementation to the polarization and politicization of cognitions and the inefficient use of resources under fragile settings.

Emergence of policy concepts

The chapters in this final category examine three policy concepts that are in the phase of emergence in their respective case contexts, namely 'climate change adaptation' (Chapter 12), 'adaptive policy responses' (Chapter 13) and 'Building with Nature' (Chapter 14). In the preceding sections, we concluded that the observed transfers of established policy knowledge and recognized international concepts were initiated or introduced as such by groups and organizations in the developed world. Similarly, the internal implementation of emerging policy concepts often starts in developed countries, as Chapters 12 to 14 demonstrate. Given the current ease and frequency with which information and ideas cross borders, these cases of emergence are unlikely to have developed completely independently from similar experiences in other areas. The difference thus between these cases and the cases discussed thus far, is the low degree to which these policy concepts are being deliberately transferred as one concept, from one case to another, in a more or less organized manner.

The Netherlands case explores how climate change adaptation is transferred from the national level to the local level and how the concept is interpreted at the local level. The motivation to implement climate adaptation measures differs based on the level of abstraction. From the national perspective, the efforts are generally related to a high level of awareness of the future issues expected to occur and the high level of associated cost and risk. This relatively voluntary response is then translated to the lower administrative (municipal) levels through slightly more coercive means. The municipalities must then incorporate the national goals alongside their more specific local goals in order to remain credible. In the analysed case, municipalities also appear to have a sense of peer pressure which further encourages similarly adaptive actions. In Chapter 13, the authors examine the way in which governmental actors deal with and respond to problems of water scarcity in an adaptive way. Experiences from two southern European countries (Spain and France) are compared with those from two northern European countries (the United Kingdom and the Netherlands). In all countries, the general water governance regime is observed as undergoing change with respect to how water scarcity is being addressed. A variety of contextual factors accompany these changes and there is also variation in the types of changes that are developing. The role of the public and their perception of the state and urgency of various contextual factors was seen to be important in determining how adaptive the policy responses were and how much they were able to move against existing norms with respect to private property rights and problem responsibility. When there was a relatively low sense of urgency experienced by the public and technological solutions were seen as being capable of solving future issues, most changes were voluntary and maintained the status quo. This was different in cases where a higher sense of urgency was held by the stakeholders, which often accompanied more coercive (formal and informal) types of processes that would include new stakeholders, new problem perspectives and thus new recognized rivalries. With empirical data from various European Union (EU) countries, Chapter 14 examines two transfers, namely the actual transfer of the EU Natura 2000 biodiversity policy from the supranational level to the member-state level, and the prospective transfer of 'Building with Nature', an emerging approach to water infrastructure design and development, to projects in the EU member states. Of the various cases explored in this book, this chapter presents the clearest examples of 'institutional isomorphism' (see Chapter 2 for an explanation). The EU Birds and Habitats Directives oblige member states to designate Natura 2000 areas and to manage these areas accordingly. As a result of the experiences with Building with Nature, this approach is increasingly recognized as a 'best practice' and is therefore becoming a legitimized approach in the development of water infrastructure projects.

Regarding the role of contextual factors in the transfer of emerging concepts, the various chapters reach multiple conclusions. Chapter 12 addresses how climate adaptation issues in the Netherlands are translated through a set of different storylines in order to be implemented at the local level. The underlying storyline that has enabled the effective translation of these policies is related to water management and its related concerns. Good water management has maintained itself as a consistently valued and understood policy goal that has coincided with efforts designed to address climate adaptation actions. The chapter shows that how policy concepts emerge closely relates to case-specific conditions, including the level of urbanization, the presence or lack of flooding experience, and the general level of the risk from flooding. Chapter 13 concludes that adaptive policy responses to water scarcity can develop based on a cognizant transfer of concepts from one region to another or as a learned response to a series of threats accompanied by the recognition of the expected continued experience of future threats. In Chapter 14, the authors identify specific (site-related) and structural (governance-related) contextual factors that affect the implementation of Natura 2000 in EU member states. By utilizing these factors, the authors conduct a prospective transfer of Building with Nature and conclude that Building with Nature stands a considerable chance of becoming a 'best practice' for water management in estuaries and coastal zones.

In all of these cases there were internal case variables that encouraged the actors to develop their water governance approaches collaboratively and according to self-reinforcing and emerging principles that fit both the present contexts and were expected to be able to achieve the desired results or address the problems that were being experienced or expected. These self-reinforcing factors were in all cases coherent with the majority of water governance panaceas discussed in Chapter 2 of this book. This is perhaps a result of the countries having similar wider contextual variables however it may also say something about the naturally appropriate manner in which to govern water. It is beyond the scope of this book to provide an adequate response to this issue, yet it is one that certainly begs further research and attention.

15.3 Reflections on the objective and approach

The initiative for this book resulted from our desire to develop a better understanding of contextual water management. As editors, we chose to adopt a broad definition of policy transfer and decided not to impose a comparative framework. Instead, we asked authors to reflect on several questions. We further promoted the use of the theoretical concepts that were introduced in Chapters 2 and 3. As the chapters evolved, we started to recognize the many forms that a policy transfer in a water management context can take. In the empirical cases, we found examples of established knowledge being transferred as well as international concepts being applied and emerging. Thus, the approach used helped us to understand the role of policy transfer in a water management context, and to discover that transfers are not only worth studying when knowledge has proven to work but also when concepts are emerging.

In analysing these different transfer processes, we have learned that transfers are unlikely to result from a strict desire of actors to learn from other actors' experiences or from external imposition or pressures. In fact, none of the authors came up with examples of transfers that were strictly voluntary or coercive. Interestingly, the desire of countries to export knowledge – a cause which is seldom mentioned in the literature – played an important role in several cases. More generally speaking, we find the transfers were shaped by external *and* internal factors. In many of the cases, a top-down or external imposition of the transfer was insufficient to effectively realize the transfer. To actually transfer knowledge, a bottom-up, or at least internal, stimulus from the receiving actors was required. Local actors, who realize the relevance of the international concepts, are the ones who shape the actual transfer. The role of external actors, such as international organizations or consultants from more developed countries, was rather to support the process or to conceptualize an externally invented concept or idea. We asked authors to describe not only the transfer process but also to reflect upon questions about contextual factors and the success of the transfer. In their assessments of their respective transfers, none of the authors chose to focus on criteria that are often mentioned in policy transfer literature, such as the degree of policy learning or the gradations of transfer (e.g. was knowledge copied or rather a source of inspiration). In the cases in which established knowledge was transferred, authors came up with specific criteria to measure effectiveness (Chapter 4) and collaboration (Chapter 5). In the cases focusing on international concepts, questions about the success of a transfer appear to be closely related to questions about contextual factors. In the chapters about emerging concepts, the focus is on contextual factors and the potential success of a transfer was not questioned.

In various chapters, authors used the Contextual Interaction Theory (presented in Chapter 3) as a theoretical lens to examine and assess policy transfers. Compared to the policy transfer literature, the theory provides a compact model that incorporates both actor characteristics and contextual factors. In various empirical chapters, the theory provided an adequate theoretical lens to explain the causes, process and outcomes of a transfer through the analysis of what motivates these actors, what information they hold to be true and how resources are distributed. At the process level, the variables of the theory help to understand the type of cooperation that emerges between actors (including the likelihood of implementation). The layers of context – often in combination with actor characteristics – are used by various actors to analyse how contextual factors impact transferability. Thus, the Contextual Interaction Theory provides a useful addition to the literature on policy transfer. Moreover, the analyses of this book provide a useful basis for systematic research regarding transfers in a water management context.

15.4 Lessons learned about contextual water management

The empirical findings of the chapters in this book support the development of a set of valuable lessons that demonstrate the relevance and importance of contextual water management. Based on the empirical findings, this section presents these lessons as our overall conclusions.

Throughout the book, the importance of various aspects of the pre-existing context in the 'receiving' countries was demonstrated over and over again. Context factors impact not only the likelihood that a certain policy idea is transferred at all, but are also of paramount importance to their degree of ultimate success. With the layers of context specified in the Contextual Interaction Theory (see Chapter 3), these factors can be systematically identified and assessed. What is more important however, is answering the question 'what main lessons can be drawn from this'?

Contextual differences provide both opportunities and constraints

It is tempting to exchange one rather simplistic axiom, such as 'internationally esteemed approaches and best practices are improvements that should be followed', by another one that sounds compelling but might preclude options for water management improvements that would have been possible. Such an axiom could be 'be aware of contextual factors in advance of the transfer to see what could fit and what will not likely fit'. The problem with such a short summary of the results presented in this book is that it can lead to a conservative bias. It is by now quite clear that policy options that are too far from the pre-existing context to be able to be cohesively incorporated will probably not be introduced, or introduced in such a way that the real meaning of the concept gets lost. It is however also quite apparent that a transfer across contexts that are extremely similar would probably not be considered to have the characteristics of a real innovation. Thus, transfers needs congruence of context, yet only congruence with the pre-existing context does not offer much potential for learning and improvement. Such conservative lessons that are drawn from the importance of contextual factors also underestimate the diversity of the people and organizations in the receiving countries and uses an evaluation period which is insufficiently long. Take, for example, the desired transfer of the 'Room for the River' approach from the Netherlands to Romania that is discussed in Chapter 4. The project was not very effective in the sense that it did not directly contribute to solving any of the flood risk problems. However, as a result of the projects that were undertaken some of the Romanian actors became more aware of this concept and are now more open to the concept than they had been before. Through these types of influences, potential foundations for change at a later time are established, perhaps even a change that evolves internally.

What then is an appropriate lesson to be drawn from the overwhelming evidence in our book that contextual factors are extremely important in policy transfers related to water management? For convenience let us refer to the Romanian example again: should actors who invested in the project conclude that it was a failure when immediate success was not achieved? Or should they have more realistic expectations, together with a longer-term orientation in which the gradual opening of mindsets (motivations, cognitions) and building of support (resources, capacity, power) is kept in focus? Of course the answer depends very much on what the transferring actors would themselves like to achieve, yet awareness of the context could lead to a more sophisticated array of responses to decisions about the continuation of a transfer project than to either just 'stay or leave'. A similar line of reasoning holds when there is no clear 'donor' and it is more a matter of ideas that are taken from the international discourse on water management (sometimes with some pressure from international organizations). Here, typically it is domestic actors that are transferring, or translating, the concepts into their own practice. Again they could base their decisions upon 'an all or nothing' perspective and with an outlook based on the achievement of immediate results, or alternatively they could orient their actions towards gradually developing a basis for further innovation. In the Mexican case, the river basin council was very outspoken in their vision that their long-term approach was the only sensible manner for achieving their goals. They focused on building a joint vision, opening the minds of the people to the importance of water issues, and increasing the acceptance of a new agency (with different geographical boundaries) and new actors. Shorter-term successes were also recognized as important in being able to achieve these longer-term goals.

Receptivity goes both ways

Continuously trying to understand the relevant contextual factors and their developments and being alert to the opportunities and threats that they pose for the potential innovation of water management practices requires a lot from the 'donor' organizations. In Chapter 3 we refer to the concept of 'receptivity' to indicate what kind of quality these people and organizations should possess. Receptivity is not only dependent on the degree of exposure to knowledge, but also more specifically on the way the actor can associate and exploit such knowledge around existing knowledge, activities and objectives (Jeffrey and Seaton, 2004). This requires that the actor 'lets the outside come in', and can open and regroup their understandings to include the new knowledge. It is also important to note that receptivity is not a passive quality but one that enables innovative and adaptive action by self-confident actors and organizations (de Boer, 2012). This is certainly an important factor for the actors on the receiving end of the policy transfer. Our considerations above allow us to conclude that in fact it is equally important for both ends of the transfer, since the donor also needs to be aware of contextual factors and their consequences and ultimately reckon with these in their actions (Bressers, 2011).

All in all, the importance of contextual factors is not an issue that should either spur or halt the transfer of innovative water management concepts. Rather, for both situations mentioned above, the lesson to be drawn can be synthesized as: 'be aware of contextual factors in order to expose where the potential misfits can occur and try to assess where the options for gradual acceptance and implementation of relevant new ideas are best placed'.

15.5 Implications and outlook

Not all policy transfers described in this book have clear donor–recipient relations. The concepts that are transferred are also often taken from the international discourse on water management and/or simultaneously stimulated by an array of organizations operating at various scales. While the top-down role of international organizations like the World Bank has a high visibility in such processes, there are increasing opportunities for other forms of transfer of water policy and management concepts that stem from other directions of learning: horizontal or even bottom-up.

Additionally, learning is increasingly taking place across different levels and scales due to the ease of global communication and hence a local action group can refer directly to internationally accepted concepts when striving for instance for an increase in participation rights. Similarly, supranational authorities such as the European Union can derive its new policy ideas directly from subnational initiatives.

Policy transfer is thus not only a subject of top-down transfers or transfers from one country to another, but in fact occurs in many different directions that sometimes skip many intermediate levels and scales. Across the globe various levels of authorities and organizations can easily learn about each other's activities and establish direct linkages to coordinate them. Global communications and networks enable policy transfers of concepts directly from the global debate to the local context and then presents its results back to the entire world. As such, more transfer – not less – is likely to be part of the future of water management.

The cases included in this book cover a comprehensive set of policy transfers that have been occurring or will soon occur in water management. It can be expected that policy transfers in water management can take even a wider range of forms. Our findings demonstrate that a multitude of contextual and actor-related factors influence the effectiveness of transfer. The overall conclusions of this book can provide the basis to develop a systematic approach to conduct comparative research regarding policy transfers in water management. We hope that the book stimulates actors with a role in the management and governance of water resources to engage and to carefully consider the international transfer of lessons.

References

Bressers, N. (2011) 'Co-creating innovation: A systemic learning evaluation of knowledge and innovation programmes', PhD thesis, Erasmus University Rotterdam.

de Boer, C. (2012) 'Contextual water management: A study of governance and implementation processes in local stream restoration projects', PhD thesis, University of Twente.

Jeffrey, P. and Seaton, R.A.F. (2004) 'A conceptual model of "receptivity" applied to the design and deployment of water policy mechanisms', *Environmental Sciences*, vol. 1, no. 3, pp. 277–300.

Index

Please note that page references to non-textual content such as Figures or Tables will be in *italics*, while the letter 'n' will follow references to Notes.

www.ingramcontent.com/pod-product-compliance
Lightning Source LLC
LaVergne TN
LVHW011115300926
849160LV00008B/396
9781138573895